THE DEVIL IN THE DETAILS

OXFORD STUDIES IN PHILOSOPHY OF SCIENCE
General Editor: Paul Humphreys, University of Virginia

The Book of Evidence
Peter Achinstein

Science, Truth, and Democracy
Philip Kitcher

The Devil in the Details: Asymptotic Reasoning in Explanation, Reduction, and Emergence
Robert W. Batterman

THE DEVIL IN THE DETAILS

Asymptotic Reasoning in Explanation,
Reduction, and Emergence

Robert W. Batterman

OXFORD
UNIVERSITY PRESS

2002

OXFORD
UNIVERSITY PRESS

Oxford New York
Athens Auckland Bangkok Bogotá Buenos Aires Cape Town
Chennai Dar es Salaam Delhi Florence Hong Kong Istanbul Karachi
Kolkata Kuala Lumpur Madrid Melbourne Mexico City Mumbai Nairobi
Paris São Paulo Shanghai Singapore Taipei Tokyo Toronto Warsaw

and associated companies in
Berlin Ibadan

Copyright © 2002 by Oxford University Press, Inc.

Published by Oxford University Press, Inc.
198 Madison Avenue, New York, New York 10016

Oxford is a registered trademark of Oxford University Press

All rights reserved. No part of this publication may be reproduced,
stored in a retrieval system, or transmitted, in any form or by any means,
electronic, mechanical, photocopying, recording, or otherwise,
without the prior permission of Oxford University Press.

Library of Congress Cataloging-in-Publication Data
Batterman, Robert W.
The devil in the details : asymptotic reasoning in explanation, reduction,
and emergence / Robert W. Batterman
p. cm.
ISBN 0-19-514647-6
1. Reasoning. 2. Science—Philosophy.
3. Science—Methodology. I. Title.
Q175.32.R45 B37 2002
153.4'3–dc21   2001047453

1 3 5 7 9 8 6 4 2

Printed in the United States of America
on acid-free paper

For Carolyn

Acknowledgments

This book is the result of many years of trying to understand certain aspects of the relations that obtain between distinct physical theories. Much of my research has been guided by the clear and fascinating work of Professor Sir Michael Berry. As is clear to anyone who reads this book, my intellectual debt to Professor Berry is very great indeed.

I also owe much to my teacher Lawrence Sklar, who taught me, among many other things, how to approach philosophical issues in the foundations of physical theory. My aim, unfortunately only asymptotically realized, has been to apply this knowledge to gain insight into some of these philosophical debates.

Mark Wilson, too, has been a great help and a source of encouragement. I want to especially thank him for many discussions that have helped me sharpen my presentation of the issues considered in the book.

Many people have aided me in my research. I wish to thank William Wimsatt for many discussions about intertheoretic reduction. Roger Jones has helped me get clear about many of the issues discussed in this book. William Taschek and Diana Raffman listened to my ramblings about multiple realizability and universality for so long that it is difficult to express sufficiently my gratitude for their tremendous help and patience. Justin D'Arms was of great help in getting clear about many aspects of the arguments that appear in this book. Sylvia Berryman has been a superb colleague. She read and provided valuable comments on virtually every chapter of the book. I would also like to thank Daniel Farrell, Dion Scott-Kakures, Joseph Mendola, George Pappas, and Alex Rueger for helpful comments and discussions, both substantive and strategic, about the writing of this book.

Several anonymous referees provided valuable criticism and commentary. I hope that I have been able to address at least some of their worries.

I would also like to acknowledge the generous support of the National Science Foundation for several grants that allowed me to pursue the research for this book.

Some of the material in this book has been taken from previously published papers. I wish to acknowledge Blackwell Publishers for permission to use material from "Explanatory Instability," *Nous, 36*, pp. 325–348, (1992). Similar thanks to Kluwer Academic Publishers for permission to use material from "Theories Between Theories ...," *Synthese, 103*, pp. 171–201, (1995) and to Oxford University Press for permission to use of material from "Multiple Realizability and Universality," *British Journal for the Philosophy of Science, 51*, pp. 115–145, (2000).

Finally, I would like to thank Carolyn for once glancing accidentally at a page of this book, and for giving me the title. I also want to thank Monty for taking me for walks to clear my head.

Contents

1	**Introduction**	**3**
2	**Asymptotic Reasoning**	**9**
	2.1 The Euler Strut	9
	2.2 Universality	13
	2.3 Intertheoretic Relations	17
	2.4 Emergence	19
	2.5 Conclusion	22
3	**Philosophical Theories of Explanation**	**23**
	3.1 Different Why-Questions	23
	3.2 Hempelian Explanation and Its Successors	25
	3.3 Conclusion	35
4	**Asymptotic Explanation**	**37**
	4.1 The Renormalization Group (RG)	37
	4.2 The General Strategy	42
	4.3 "Intermediate Asymptotics"	44
	4.4 Conclusion: The Role of Stability	57
5	**Philosophical Models of Reduction**	**61**
	5.1 Nagelian Reduction	62
	5.2 Multiple Realizability	65
	5.3 Kim's "Functional Model of Reduction"	68
	5.4 A Metaphysical Mystery	71
	5.5 Multiple Realizability as Universality	73
	5.6 Conclusion	76
6	**Intertheoretic Relations—Optics**	**77**
	6.1 "Reduction$_2$"	78
	6.2 Singular Limits	80
	6.3 Wave and Ray Theories	81
	6.4 Universality: Diffraction Catastrophe Scaling Laws	93

	6.5 Conclusion	95
7	**Intertheoretic Relations—Mechanics**	**99**
	7.1 Classical and Quantum Theories	100
	7.2 The WKB Method	104
	7.3 Semiclassical "Emergents"	109
	7.4 Conclusion	110
8	**Emergence**	**113**
	8.1 Emergence and the Philosophy of Mind	114
	8.2 The Rainbow Revisited: An Example of Emergence?	115
	8.3 A New Sense of Emergence	121
	8.4 Tenet 5: Novel Causal Powers?	126
	8.5 Conclusion	128
9	**Conclusions**	**131**
Bibliography		**137**
Index		**141**

THE DEVIL IN THE DETAILS

1

Introduction

Methodological philosophy of science concerns itself, among other things, with issues about the nature of scientific theories, of scientific explanation, and of intertheoretic reduction. Philosophers of science frequently have attempted to identify and "rationally reconstruct" distinct types of reasoning employed by scientists as they go about their business. Philosophical questions often asked in these contexts include: What counts as an explanation? When does one theory replace or reduce another? What, for that matter, is a theory? All too often, however, these reconstructions end up being quite far removed from the actual science being done. Much of interest remains in the details and gets lost in the process of philosophical abstraction.

Recently, though, philosophers of science have begun to provide more nuanced and scientifically better informed approaches to these types of methodological questions. I intend this discussion to be one that pays close attention to a certain type of reasoning that plays a role in understanding a wide range of physical phenomena—one that I think has largely been missed by philosophers of science both in the past and even of late. Somewhat ironically (given the last sentence of the last paragraph), this type of reasoning involves, at its heart, a type of abstraction—a means for ignoring or throwing away various details. It is, though, a type of reasoning that is motivated from within the scientific enterprise, and not, as old-style rational reconstructions of scientific reasoning, motivated by external philosophical programs and prejudices. I call this kind of reasoning "asymptotic reasoning," and I hope to show how crucial it is to the scientific understanding of many aspects of physical phenomena. Once this is properly recognized, it will inform our understanding of many aspects of scientific methodology.

The idea that scientific understanding often requires methods which eliminate detail and, in some sense, precision, is a theme that runs throughout this book. Suppose we are interested in explaining some physical phenomenon governed by a particular physical theory. That theory may say a lot about the nature of the phenomenon: the nature of its evolution, and what sorts of details—for example, initial and boundary conditions—are required to "solve" the governing equations, and so on. One might think that the theory will therefore enable us to account for the phenomenon through straightforward derivation from the appropriate initial data, given the governing equation(s).[1] For some types of why-questions this may very well be the case. However, I will show that, with respect to other critically important why-questions, *many theories are explanatorily deficient*. This is true even for those theories that are currently taken to be so wellconfirmed as to constitute paradigms of scientific achievement.

The kind of explanatory questions for which the detailed accounts simply provide explanatory "noise" and for which asymptotic methods fill in the explanatory lacunae are questions about the existence of patterns noted in nature. Details are required to account for why a given instance of a pattern can arise, but such details obscure and even block understanding of why the pattern itself exists. Physicists have a technical term for these patterns of behavior. They call them "universal." Many systems exhibit similar or identical behavior despite the fact that they are, at base, physically quite distinct. This is the essence of universality. Examples abound in the literature on the thermodynamics of phase transitions and critical phenomena. Such wildly diverse systems as fluids and magnets exhibit the same behavior when they are in certain critical states. Asymptotic methods such as the renormalization group provide explanations for this remarkable fact. They do so by providing *principled reasons* grounded in the fundamental physics of the systems for why many of the details that genuinely distinguish such systems from one another are *irrelevant* when it comes to the universal behavior of interest.

While most discussions of universality and its explanation take place in the context of thermodynamics and statistical mechanics, we will see that universal behavior is really ubiquitous in science. Virtually any time one wants to explain some "upper level" generalization, one is trying to explain a universal pattern of behavior. Thus, this type of explanatory strategy—what I call "asymptotic explanation"—should play a role in the various philosophical debates about the status of the so-called special sciences. I will argue that the infamous multiple realizability arguments that feature prominently in these discussions are best understood in the context of trying to explain universal behavior. Multiple realizability is the idea that there can be heterogeneous or diverse "realizers" of "upper level" properties and generalizations. But this is just to say that those upper level properties and the generalizations that involve them—the "laws" of the special sciences—are universal. They characterize similar behavior in physically distinct systems.

The reference here to the status of the special sciences will immediately call to mind questions about relationships between distinct theories. If, for instance, psychology is a genuine science, what is its relationship to physics? So-called nonreductive physicalists want to maintain the *irreducibility* of the science of the mental to more fundamental physical theory, while at the same time holding on to the idea that at base there is nothing other than physics—that is, they

---

[1]This idea is at the center of many extant conceptions of scientific explanation.

Introduction

maintain that we don't need to reify mental properties as ontologically distinct from physical properties. This is one member of a truly thorny set of issues. The debates in the literature focus largely on questions about the reduction of one theory to another. I will argue that this problematic needs rethinking. Questions about reduction—what is its nature, and whether it is possible at all—are much more subtle than they are often taken to be.

Understanding the nature of intertheoretic reduction is, surely, an important topic in methodological philosophy of science. But most of the literature on reduction suffers, I claim, from a failure to pay sufficient attention to detailed features of the respective theories and their interrelations. Those cases for which something like the philosophers' (Nagelian or neo-Nagelian) models of reduction will work are actually quite special. The vast majority of purported intertheoretic reductions, in fact, fail to be cases of reduction. It is best to think about this in the context of a distinction between types of reductions recognized first by Thomas Nickles (1973). On the usual philosophical models, a typically newer, more refined theory, such as quantum mechanics, is said to reduce a typically older, and coarser theory, such as classical mechanics. Thus, classical mechanics is said to *reduce to* quantum mechanics. On the other hand, Nickles noted that physicists often speak of the reduction relation as the inverse of this. They hold that the more refined theory reduces to the coarser theory in some sort of correspondence limit. Thus, on this view, quantum mechanics is supposed to *reduce to* classical mechanics in an appropriate limit. The so-called correspondence principle in quantum mechanics is a paradigm example of this type of reductive limit. Somehow, quantum mechanics is supposed to "go over into" classical mechanics in some limit as "things get big" or, perhaps, as Planck's constant approaches a limiting value.

However, there are deep and subtle problems here. These limiting relations can be of two sorts. Roughly, some theories will "smoothly approach" another in a relevant correspondence limit. For other theory pairs, the limit can be singular. This means that the behavior at the limit is fundamentally different from the behavior as the limit is being approached. I think that a case can be made that philosophical models of reduction will apply only if the limiting relation between the theory pairs is smooth or regular. Thus, any hope for a philosophical reduction will depend on the satisfaction of the "physicists' " limiting relation. If the relationship is singular, however, things are much more complicated. In fact, I will argue that this is an indication that *no reduction* of any sort can obtain between the theories.

Despite the failure of reductive relations between some theories, much of interest, both physical and philosophical, can be gained by studying the asymptotic behavior of theories in these singular correspondence limits. I will discuss several examples of this throughout the book. One that will receive a lot of attention is the relationship between the wave and ray theories of light. A specific example here will occupy much of our attention. This is the example of the rainbow. Certain features of rainbows can be fully understood only through asymptotic methods. In effect, these are *universal* features that "emerge" in the asymptotic domain as the wave theory approaches the ray theory in the limit as the wavelength of light approaches zero. They inhabit (to speak somewhat metaphorically) an asymptotic borderland between theories. I will argue that a third explanatory theory is required for this asymptotic domain. The phenomena inhabiting this borderland are not explainable in purely wave theoretic or ray theoretic terms. The accounts required to characterize and explain these borderland phenomena deserve the title "theory." In part, this is because the fundamental wave theory is explanatorily deficient. As we will see, the theory of the borderland incorporates, in well-defined ways, features of both the wave and ray theories. Asymptotic reasoning plays a key explanatory and interpretive role here.

In general, when asymptotic relations between theories are singular, we can expect such "no man's lands" where new phenomena exist and where new explanatory theories are required. This talk of "new phenomena" and the necessity of "new theories," together with my use of the term "emergent" earlier suggests that asymptotic investigations may also inform various philosophical debates about the nature and status of so-called emergent properties. I will argue that this is indeed the case.

One important aspect of the "received" opinion about emergent properties is that they are best understood in mereological—part/whole—terms: A property of a whole is emergent if it somehow transcends the properties of its parts. Furthermore, symptoms of this type of transcendence include the unexplainability and unpredictability of the emergent features from some underlying, "base" or fundamental theory. Likewise, the received view holds that the emergent properties are irreducible to that base theory. My discussion of the asymptotic nature of the new phenomena and new theories will lead to a different understanding of emergent properties. Part/whole relations will turn out to be inessential, or unnecessary, for emergence. Some phenomena for which no part/whole relations are discernible must reasonably be considered emergent. What is essential is the singular nature of the limiting relations between the "base" theory and the theory describing the emergents. The singular nature of this limiting relationship is, as just noted, the feature responsible for the failure of the physicists' conception of reduction. Emergence depends, therefore, on a failure of reducibility. This clearly fits with the received view, although the proper understanding of reduction is, as I have suggested, distinct from most of the "standard" views.

We will see, however, that while reductive failure of a certain type is necessary for emergence, it does not entail (as it is typically taken to) the necessary failure of explainability. Emergent properties are universal. It is legitimate to search for, and expect, explanations of their universality. Contrary to received opinion, such properties are not brute and inexplicable features of the world. As we will see in several places throughout the book, reduction and explanation, when properly understood, do not march in lock-step. Asymptotic explanations are possible even for phenomena that are in an important sense irreducible and emergent.

The preceding discussion is intended to give the reader a brief indication of the various topics considered in the following pages. The unifying theme, as

I've indicated, is the role played by asymptotic reasoning. The claim is that by focusing on this ubiquitous form of scientific reasoning, new insights can be gained into old philosophical problems. Thus, speaking negatively, I hope to show that (a) philosophers of science have, by and large, missed an important sense of explanation, (b) extant philosophical accounts of emergence must be refined in various ways, and (c) issues about reductive relations between theories are much more involved than they are typically taken to be. More positively, we will see that asymptotic reasoning leads to better informed accounts of at least certain aspects of explanation, reduction, and emergence.

It is important here, I think, to say a few things about the nature and scope of the discussions to follow. I do not intend to provide detailed discussions of the many different accounts of explanation, reduction, and so on that appear in the philosophical literature. Instead, I will concentrate on providing a positive proposal, motivated through an examination of what I take to be representative or core positions.

As a result, this is a short book. It also involves fairly technical discussions in some places. As I've already noted, the key to understanding the importance of asymptotic reasoning is to examine in some detail certain examples in which it is used. These are, by their very nature, described in mathematical terms. Asymptotic methods, in mathematical physics and in the mathematics of the applied sciences, have only recently received clear and systematic formulations. Nevertheless, I believe that such methods (broadly construed) are far more widespread in the history of science than is commonly realized. They play important roles in many less technical contexts. However, to best understand these methods, it is necessary to investigate the more technical arguments appearing in recent work by physicists and applied mathematicians. It is my hope, though, that even the reader who skims the technical discussions will be able to get a tolerably clear idea of how these methods are supposed to work.

The devil is truly in the details. And, even though the aim is to understand the importance of systematic methods for throwing details away, this understanding is achievable only through fairly close examinations of specific examples.

2

Asymptotic Reasoning

This chapter will introduce, via the consideration of several simple examples, the nature and importance of asymptotic reasoning. It is necessary that we also discuss an important feature of many patterns or regularities that we may wish to understand. This is their universality. "Universality," as I've noted, is the technical term for an everyday feature of the world—namely, that in certain circumstances distinct types of systems exhibit similar behaviors. (This can be as simple as the fact that pendulums of very different microstructural constitutions all have periods proportional to the square root of their length. See section 2.2.) We will begin to see why asymptotic reasoning is crucial to understanding how universality can arise. In addition, this chapter will begin to address the importance of asymptotics for understanding relations between theories, as well as for understanding the possibility of emergent properties. Later chapters will address all of these roles and features of asymptotic reasoning in more detail.

2.1 The Euler Strut

Let us suppose that we are confronted with the following physical phenomenon. A stiff ribbon of steel—a strut—is securely mounted on the floor in front of us. Someone begins to load this strut symmetrically. At some point, after a sufficient amount of weight has been added, the strut buckles to the left. See figure 2.1. How are we to understand and explain what we have just witnessed?

Here is an outline of one response. At some point in the weighting process (likely just prior to the collapse), the strut reached a state of unstable equilibrium called the "Euler critical point." This is analogous to the state of a pencil balancing on its sharpened tip. In this latter case, we can imagine a hypothetical situation in which there is nothing to interfere with the pencil—no breeze in the room, say. Then the pencil would presumably remain in its balanced state forever. Of course, in the actual world we know that it is very difficult to maintain such a balancing act for any appreciable length of time. Similarly,

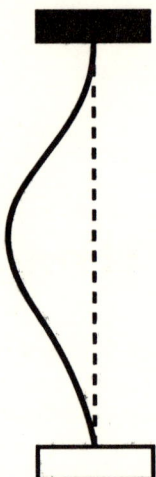

Figure 2.1: Buckling strut

molecular collisions will "cause" the strut to buckle either to the left or to the right. Either of these two buckled states is more stable than the critical state in that the addition of more weight will only cause it to sag further on the same side to which it has already collapsed.

So, in order to explain why the strut collapsed to the left, we need to give a complete causal account that (1) characterizes the details of the microstructural makeup of the particular strut, (2) refers to the fact that the strut had been weighted to the critical point, and (3) characterizes the details of the chain of molecular collisions leading up to the one water vapor molecule, the culprit, that hits the strut on its right side. If we were actually able to provide all these details, or at least some relevant portion of them, wouldn't we have an explanation of what we observed? Wouldn't we understand the phenomenon we have witnessed?

Both common sense and at least one prominent view of the nature of explanation and understanding would have it that we would now understand what we have seen. By providing this detailed causal account, we will have shown how the particular occurrence came about. We will have displayed the mechanisms which underlie the phenomenon of interest. On this view, the world is generally opaque. Providing accounts like this, however, open up "the black boxes of nature to reveal their inner workings" (Salmon, 1989, p. 182). We can call this view a causal-mechanical account.

On Peter Railton's version of the causal-mechanical account, the detailed description of the mechanisms that provides our explanation is referred to as an "ideal explanatory text."

> [A]n ideal text for the explanation of the outcome of a causal process would look something like this: an inter-connected series of

law-based accounts of all the nodes and links in the causal network culminating in the explanandum, complete with a fully detailed description of the causal mechanisms involved and theoretical derivations of all of the covering laws involved.... It would be the whole story concerning why the explanandum occurred, relative to a correct theory of the lawful dependencies of the world. (Railton, 1981, p. 247)

For the strut, as suggested, this text will refer to its instability at the critical point, to the fact that it is made of steel with such and such atomic and molecular structure, and to the details of the collision processes among the "air molecules" leading up to the buckling.

But how satisfying, actually, is this explanation? Does it really tell us the whole story about the buckling of the strut? For instance, one part of the "whole story" is how this particular account will bear on our understanding of the buckling of an "identical" strut mounted next to the first and which buckled to the right after similar loading. Was what we just witnessed a fluke, or is the phenomenon repeatable? While we cannot experiment again with the very same strut—it buckled—we still might like to know whether similar struts behave in the same way. Going a bit further, we can ask whether our original causal-mechanical story sheds any light on similar buckling behavior in a strut made out of a different substance, say, aluminum? I think that the story we have told has virtually no bearing whatsoever on these other cases. Let me explain.

Let's consider the case of a virtually identical strut mounted immediately next to the first. What explains why it buckled to the right after having been loaded just like the first one? On the view we are considering, we need to provide an ideal explanatory text, which, once again, will involve a detailed account of the microstructural make-up of *this* strut, reference to the fact that it has been loaded to its critical point, and, finally, a complete causal story of all the molecular collisions leading up to the striking on the left side by a particular dust particle. Most of these details will be completely different than in the first case. Even though both struts are made of steel, we can be sure that there will be differences in the microstructures of the two struts—details that may very well be causally relevant to their bucklings. For instance, the location of small defects or fractures in the struts will most likely be different. Clearly, the collision histories of the various "air molecules" are completely distinct in the two cases as well. After all, they involve different particles. The two explanatory texts, therefore, are by and large completely different. Had we been given the first, it would have no bearing on our explanation of the buckling of the second strut.

In the case of an aluminum strut, the explanatory texts are even more disjoint. For instance, the buckling load will be different since the struts are made of different materials. Why should our explanation of the behavior of a steel strut bear in any way upon our understanding of the behavior of one composed of aluminum?

At this point it seems reasonable to object: "Clearly these struts exhibit similar behavior. In fact, one can characterize this behavior by appeal to Euler's formula:[1]

$$P_c = \pi^2 \frac{EI}{L^2}.$$

How can you say the one account has nothing to do with the other? Part of understanding how the behavior of one strut can bear on the behavior of another is the recognition that Euler's formula applies to both." (Here P_c is the critical buckling load for the strut. The formula tells us that this load is a function of what the strut is made of as well as certain of its geometric properties—in particular, the ration I/L^2.)

I agree completely. However, the focus of the discussion has shifted in a natural way from the particular buckling of the steel strut in front of us to the understanding of buckling behavior of struts in general. These two foci are not entirely distinct. Nevertheless, nothing in the ideal explanatory text for a particular case can bear upon this question. "Microcausal" details might very well be required to determine a theoretical (as opposed to a measured phenomenological) value for Young's modulus E of the particular strut in front of us, but what, in all of these details, explains why what we are currently witnessing is a phenomenon to which Euler's formula applies? The causal-mechanical theorist will no doubt say that all of the microcausal details about this strut will yield an understanding of why *in this particular case* the Euler formula is applicable: These details will tell us that E is what it is, and when all the evidence is in, we will simply see that P is proportional to I/L^2.

But, so what? Do we *understand* the phenomenon of strut buckling once we have been given all of these details? Consider the following passage from a discussion of explanation and understanding of critical phenomena. (The technical details do not matter here. It is just important to get the drift of the main complaint.)

> The traditional approach of theoreticians, going back to the foundation of quantum mechanics, is to run to Schrödinger's equation when confronted by a problem in atomic, molecular, or solid state physics! One establishes the Hamiltonian, makes some (hopefully) sensible approximations and then proceeds to attempt to solve for the energy levels, eigenstates and so on.... The modern attitude is, rather, that the task of the theorist is to *understand* what is going on and to elucidate which are the crucial features of the problem. For instance, if it is asserted that the exponent α depends on the dimensionality, d, and on the symmetry number, n, but on no other factors, then the theorist's job is to explain *why* this is so and subject to what provisos. If one had a large enough computer to solve Schrödinger's equation and the answers came out that way, one would still have *no understanding* of why this was the case! (Fisher, 1983, pp. 46–47)

[1] E is Young's modulus characteristic of the material. I is the second moment of the strut's cross-sectional area. L is the length of the strut.

If the explanandum is the fact that struts buckle at loads given by Euler's formula, then this passage suggests, rightly I believe, that our causal-mechanical account fails completely to provide the understanding we seek. All of those details that may be relevant to the behavior of the particular strut don't serve to answer the question of why loaded struts in general behave the way that they do. Actually, what does the explaining is a systematic method for abstracting from these very details.

The point of this brief example and discussion is to motivate the idea that sometimes (actually, very often, as I will argue) *science requires methods that eliminate both detail and, in some sense, precision.* For reasons that will become clear, I call these methods "asymptotic methods" and the type(s) of reasoning they involve "asymptotic reasoning."

2.2 Universality

The discussion of Euler struts in the context of the causal-mechanical view about explanation leads us to worry about how similar behaviors can arise in systems that are composed of different materials. For instance, we have just seen that it is reasonable to ask why Euler's formula describes the buckling load of struts made of steel as well as struts made of aluminum. In part this concern arises because we care whether such a phenomenon is repeatable. Often there are pragmatic reasons for why we care. For instance, in the case of buckling struts, we may care because we intend to use such struts or things like them in the construction of buildings. But despite (and maybe because of) such pragmatic concerns, it seems that science often concerns itself with discovering and explaining similar patterns of behavior.

As I noted in chapter 1, physicists have coined a term for this type of phenomenon: "universality." Most broadly, a claim of universality is an expression of behavioral similarity in diverse systems. In Michael Berry's words, saying that a property is a "universal feature" of a system is "the slightly pretentious way in which physicists denote identical behaviour in different systems. The most familiar example of universality from physics involves thermodynamics near critical points" (Berry, 1987, p. 185).

There are two general features characteristic of universal behavior or universality.

1. The details of the system (those details that would feature in a complete causal-mechanical explanation of the system's behavior) are largely irrelevant for describing the behavior of interest.

2. Many different systems with completely different "micro" details will exhibit the identical behavior.

The first feature is, arguably, responsible for the second. Arguments involving appeal to asymptotics in various forms enable us to see how this is, in fact, so.

It is clear that we can think of the Euler formula as expressing the existence of universality in buckling behavior. The formula has essentially two components.

First, there is the system—or material—specific value for Young's modulus. And second, there are the "formal relationships" expressed in the formula.

To see how ubiquitous the concept of universality really is, let us consider another simple example. We want to understand the behavior of pendulums. Particularly, we want to understand why pendulums with bobs of different colors and different masses, rods of different lengths, often composed of different materials, all have periods (for small oscillations) that are directly proportional to the square root of the length of the rod from which the bob is hanging. In other words, we would like to understand *why* the following relation generally holds for the periods, θ, of pendulums exhibiting small oscillations:[2]

$$\theta = 2\pi \sqrt{\frac{l}{g}}. \tag{2.1}$$

One usually obtains this equation by solving a differential equation for the pendulum system. The argument can be found near the beginning of just about every elementary text on classical mechanics. In one sense this is an entirely satisfactory account. We have a theory—a well-confirmed theory at that—which through its equations tells us that the relevant features for the behavior of pendulum systems are the gravitational acceleration and the length of the bob. In a moment, we will see how it is possible to derive this relationship without any appeal to the differential equations of motion. Before getting to this, however, it is worthwhile asking a further hypothetical question. This will help us understand better the notion of universality and give us a very broad conception of asymptotic reasoning.

Why are factors such as the color of the bob and (to a large extent) its microstructural makeup irrelevant for answering our why-question about the period of the pendulum? There are many features of the bob and rod that constitute a given pendulum that are clearly irrelevant for the behavior of interest. What allows us to set these details aside as "explanatory noise"? Suppose, hypothetically, that we did not have a theory that tells us what features are relevant for specifying the state of a pendulum system. Suppose, that is, that we were trying to *develop* such a theory to explain various observed empirical regularities "from scratch," so to speak. In such a pseudo-history would a question about the relevance of the color of the bob to its period have seemed so silly? The very *development of the theory* and the differential equation that describes the behavior of pendulums involved (the probably not so systematic) *bracketing as irrelevant many of the details and features that are characteristic of individual systems.*

Next, suppose we are in a state of knowledge where we believe or can make an educated guess that the period of the pendulum's swing depends only on the mass of the bob, the length of the pendulum, and the gravitational acceleration. In other words, we know something about classical mechanics—for instance, we have progressed beyond having to worry about color as a possible variable to be

---

[2]Here "l" denotes the length of the rod and "g" is the acceleration due to gravity.

considered. Can we, without solving any differential equations—that is, without appeal to the detailed theory—determine the functional relationship expressed in equation (2.1)? The answer is yes; and we proceed to do so by engaging in dimensional analysis (Barenblatt, 1996, pp. 2–5). In effect, the guess we have just made is sufficient to answer our why-question about the period of a pendulum.

We have a set of objects that represent units of length, mass, and time. These are standards everyone agrees upon—one gram, for instance, is 1/1000 of the mass of a special standard mass in a vault in the Bureau of Weights and Measures in Paris. Given these standards we will have a system of units for length, mass, and time (where L is the dimension of length, M is the dimension of mass, and T is the dimension of time). For our pendulum problem we have guessed that only the length l of the pendulum, its mass m, and the gravitational acceleration g should be relevant to its period θ. Note that l, m, and g are numbers holding for a particular choice of a system of units of measurement (e.g., centimeters, grams, and seconds). But in some sense that choice is arbitrary. Dimensional analysis exploits this fundamental fact—namely, that *the physics should be invariant across a change of fundamental units of measurement.*

The dimensions for the quantities involved in our problem are the following:[3]

$$[\theta] = T; \ [l] = L; \ [m] = M; \ [g] = LT^{-2}.$$

Now, consider the quantity l/g. If the unit of length is decreased by a factor of a, and the unit of time is decreased by a factor of b, then the numerical value of length in the numerator increases by a factor of a and the numerical value of acceleration in the denominator increases by a factor ab^{-2}. This implies that the value of the ratio l/g increases by a factor of b^2. Hence, the numerical value of $\sqrt{l/g}$ increases by a factor of b. Since the numerical value for the period, θ, would also increase by a factor of b under this scenario (decreasing the unit of time by a factor of b), we know the quantity

$$\Pi = \frac{\theta}{\sqrt{l/g}} \quad (2.2)$$

remains invariant under a change in the fundamental units. This quantity Π is dimensionless. In the jargon of dimensional analysis, we have "nondimensionalized" the problem.

In principle, Π depends (just like θ under our guess) upon the quantities l, m, and g: $\Pi = \Pi(l, m, g)$. If we decrease the unit of mass by some factor c, of course, the numerical value for mass will increase by that same factor c. But, in so doing, neither Π nor l nor g will change in value. In particular, $\Pi(l, m, g)$ is independent of m. What happens to Π if we decrease the unit of length by some factor a leaving the unit of time unchanged? While the value for length will increase by a factor of a, the quantity Π, as it is dimensionless, remains unchanged. Hence, $\Pi(l, m, g)$ is independent of l. Finally, what happens to Π

---

[3] "$[\bullet]$" $\stackrel{\text{def}}{=}$ "the dimension of $\bullet$".

if we decrease the unit of time by a factor of b while leaving the unit of length invariant. We have seen that this results in the numerical value for acceleration g increasing by a factor of b^{-2}. However, Π and l, and m remain unchanged.

This establishes the fact that $\Pi(l, m, g)$ is independent of all of its parameters. This is possible only if Π is a constant:

$$\Pi = \frac{\theta}{\sqrt{l/g}} = constant. \tag{2.3}$$

Hence,

$$\theta = constant \sqrt{\frac{l}{g}}, \tag{2.4}$$

which, apart from a constant, just is equation (2.1). The constant in (2.4) can be easily determined by a single measurement of the period of oscillation of a simple pendulum.

This is indeed a remarkable result. To quote Barenblatt: "[I]t would seem that we have succeeded in obtaining an answer to an interesting problem from nothing—or, more precisely, only from a list of the quantities on which the period of oscillation of the pendulum is expected to depend, and a comparison (analysis) of their dimensions" (Barenblatt, 1996, p. 5). No details whatsoever about the nature of individual pendulums, what they are made of, and so on, played any role in obtaining the solution.

This example is really a special (degenerate) case. In most problems, the equation for the dimensionless quantity of interest, Π, that results from the analysis will not equal a constant, but rather will be a function of some other dimensionless parameters:

$$\Pi = \Phi(\Pi_1, \ldots, \Pi_m).$$

In such cases, the analysis proceeds by trying to motivate the possibility that one or more of the Π_i's can be considered extremely small or extremely large. Then one can further reduce the problem by taking a limit so that the Π_i can be replaced by a constant: $\Pi_i(0) = C$ or $\Pi_i(\infty) = C$. This would yield an equation

$$\Pi = \Phi(\Pi_1, \ldots, \Pi_{i-1}, C, \Pi_{i+1}, \Pi_m), \tag{2.5}$$

which, one hopes, can be more easily solved. This recipe, however, involves a strong assumption—one that is most often false. This is the assumption that the limits $\Pi_i(0)$ or $\Pi_i(\infty)$ actually exist. As we will see, when they do not, dimensional analysis fails and interesting physics and mathematics often come into play.

The appeal to limiting cases, whether regular (where the limits $\Pi_i(0)$ or $\Pi_i(\infty)$ do exist) or singular (where those limits fail to exist), constitutes paradigm instances of asymptotic reasoning. We will see many examples of such reasoning later. The important point to note here has to do with the relationship between universality and asymptotic reasoning of this sort. It is often the case that the result of this kind of reasoning about a given problem is the discovery of some relationship like (2.3) or, more generally, like (2.5). In other words, asymptotic analysis often leads to equations describing universal features of systems. This happens because *these methods systematically eliminate irrelevant details about individual systems.*

Before leaving this section on universality, let me try to forestall a particular misunderstanding of what universality is supposed to be. As the discussion of the "pseudo history" of the pendulum theory is meant to show, it is not only a feature of highly technical physical phenomena. Universality as expressed in 1 and 2 holds of everyday phenomena in science. While many everyday patterns exhibit universality, they are not, therefore, mundane. The simple observable fact that systems at different temperatures tend toward a common temperature when allowed to interact with each other is an everyday occurrence. As an instance, just think of a glass of ice water coming to room temperature. Surely this regularity is universal—whether we consider ice or a rock at a cooler temperature interacting with the warmer room, the same pattern is observed. One should not be misled by the everyday nature of this pattern into thinking that the explanation for the pattern is at all trivial. Deep results in statistical mechanics involving asymptotics are necessary to explain this phenomenon. (This particular explanation will be discussed in section 8.3.) I think that despite the everyday occurrences of universal behavior, philosophers of science have not, by and large, understood how such patterns and regularities are to be explained. I will have much more to say about this type of explanation in chapter 4 and elsewhere throughout the book.

Let's turn now to a brief discussion of what may seem to be a completely different topic—theoretical reduction. As we will see, however, there are intimate connections between questions about the explanation of universality and the various ways different theories may be related to one another.

2.3   Intertheoretic Relations

Philosophers of science have always been concerned with how theories of one "domain" fit together with theories of some other. Paradigm examples from physics are the relations obtaining between classical thermodynamics and statistical mechanics; between the Newtonian physics of space and time and the "theory" of relativity; between classical mechanics and quantum mechanics; and between the ray theory of light and the wave theory. Most philosophical discussions of these interrelations have been framed in the context of questions about reduction. Do (and if so how) the former members of these pairs reduce to the latter members? "Reduction," here, is typically understood as some variant of the following prescription:

> A theory T' reduces to a theory T if the *laws* of T' are derivable (in some sense) from those of T.

This conception of reduction may require the identification (or nomic correlation) of properties in the reduced theory (T') with those in the reducing (T). With this requirement may come all sorts of difficulties, both conceptual and technical in nature. Much also will depend upon what sort of derivability is required and how strict *its* requirements are. These are all details that I think we can safely avoid at this point in the discussion. Most philosophical accounts of reduction along these lines also require that the reducing theory *explain* the reduced theory—or at least explain why it works as well as it does in its domain of applicability.

For example, the special theory of relativity is supposed to reduce Newtonian space and time. This reduction is accomplished by "deriving" Newtonian laws where the derivation involves some kind of limiting procedure. In particular, one shows that the Newtonian theory "results" when velocities are slow compared with the speed of light. In so doing the reduction also is supposed to explain the (approximate) truth of the Newtonian conception of space and time. It explains why, for instance, we can think of space and time as divided unambiguously into spaces at distinct times, even though according to special relativity the conception of absolute simultaneity required for this conception does not exist. Clearly questions about the relationships between properties and concepts in the two theories will immediately come to the fore.

Physicists, typically, use the term "reduction" in a different way than do philosophers.[4] For each of the theory pairs mentioned in the earlier paragraph, the physicist would speak of the second member of the pair reducing to the first. Reduced and reducing theories are inverted in comparison to the philosophers' way of understanding reduction. For example, physicists refer to the "fact" that quantum mechanics reduces to classical mechanics in some kind of correspondence limit—say where we let Planck's constant $\hbar \to 0$.[5] Another example is the reduction of special relativity to Newtonian space and time in the limit $(v/c)^2 \to 0$.

In general this other sense of reduction has it that a "more refined," more encompassing (and typically more recent) theory, T_f, corresponds to a "coarser," less encompassing (and typically earlier) theory, T_c, as some fundamental parameter (call it ϵ) in the finer theory approaches a limiting value. Schematically, the physicists' sense of reduction can be represented as follows:

$$\lim_{\epsilon \to 0} T_f = T_c. \qquad (2.6)$$

The equality in (2.6) can hold only if the limit is "regular." In that case, on my view, it is appropriate to call the limiting relation a "reduction." If the limit in (2.6) is singular, however, the schema fails and I think it is best to talk simply about intertheoretic relations. Let me give a brief explication of these notions.

If the solutions of the relevant formulas or equations of the theory T_f are such that for small values of ϵ they *smoothly* approach the solutions of the

---

[4]See (Nickles, 1973) for an important discussion of these different senses.

[5]There are questions about what this can really mean: How can a constant change its value? There are also questions about whether the claim of reduction is true, even once we have decided what to say about the varying "constant" problem.

corresponding formulas in T_c, then schema (2.6) will hold. For these cases we can say that the "limiting behavior" as $\epsilon \to 0$ equals the "behavior in the limit" where $\epsilon = 0$. On the other hand, if the behavior in the limit is of a *fundamentally different character* than the nearby solutions one obtains as $\epsilon \to 0$, then the schema will fail.

A nice example illustrating this distinction is the following: Consider the quadratic equation
$$x^2 + x - \epsilon 9 = 0.$$
Think of ϵ as a small expansion or perturbation parameter. The equation has two roots for any value of ϵ as $\epsilon \to 0$. In a well-defined sense, the solutions to this quadratic equation as $\epsilon \to 0$ smoothly approach the solutions to the "unperturbed" ($\epsilon = 0$) equation
$$x^2 + x = 0;$$
namely, $x = 0, -1$. On the other hand, the equation
$$\epsilon x^2 + x - 9 = 0$$
has two roots for any value of $\epsilon > 0$ but has for its "unperturbed" solution only one root; namely, $x = 9$. The equation suffers a reduction in order when $\epsilon = 0$. Thus, the character of the behavior in the limit $\epsilon = 0$ differs fundamentally from the character of its limiting behavior. Not all singular limits result from reductions in order of the equations, however. Nevertheless, these latter singular cases are much more prevalent than the former.

The distinction between regular and singular asymptotic relations is the same as that discussed in the last section between problems for which dimensional analysis works, and those for which it does not. The singular cases are generally much more interesting, both from a physical and a philosophical perspective, in that it is often the case that new physics emerges in the asymptotic regime in which the limiting value is being approached.

From this brief discussion, we can see that asymptotic reasoning plays a major role in our understanding of how various theories "fit" together to describe and explain the workings of the world. In fact, the study of asymptotic limits is part and parcel of intertheoretic relations. One can learn much more about the nature of various theories by studying these asymptotic limits than by investigating reductive relations according to standard philosophical models. This point of view will be defended in chapters 6 and 7.

2.4 Emergence

Questions about reduction and the (im)possibility of identifying or otherwise correlating properties in one theory with those in another are often related to questions about the possibility of emergent properties. In this section I will briefly characterize what I take to be a widely held account of the nature of emergence and indicate how I think it will need to be amended once one takes asymptotic limiting relations into account.

It is a presupposition of the "received" account of emergence that the world is organized somehow into levels. In particular, it is presupposed that entities at some one level have properties that "depend" in some sense on properties of the entities' constituent parts. Jaegwon Kim (1999, pp. 19–20) has expressed the "central doctrines of emergentism" in the following four main claims or tenets:

1. *Emergence of complex higher-level entities*: Systems with a higher-level of complexity emerge from the coming together of lower-level entities in new structural configurations.

2. *Emergence of higher-level properties*: All properties of higher-level entities arise out of the properties and relations that characterize their constituent parts. Some properties of these higher, complex systems are "emergent," and the rest merely "resultant."

3. *The unpredictability of emergent properties*: Emergent properties are not predictable from exhaustive information concerning their "basal conditions." In contrast, resultant properties are predictable from lower-level information.

4. *The unexplainability/irreducibility of emergent properties*: Emergent properties, unlike those that are merely resultant, are neither explainable nor reducible in terms of their basal conditions.

Kim also notes a fifth tenet having to do with what sort of causal role emergent properties can play in the world. Most emergentists hold that they must (to be genuinely emergent) play some novel causal role.

5. *The causal efficacy of the emergents*: Emergent properties have causal powers of their own—novel causal powers irreducible to the causal powers of their basal constituents (1999, p. 21).

It is evident from tenets 1–4 that the relation of the whole to its parts is a major component of the contemporary philosophical account of emergentism. Paul Teller (1992, p. 139), in fact, holds that this part/whole relationship is fundamental to the emergentist position: "I take the naked emergentist intuition to be that an emergent property of a whole somehow 'transcends' the properties of the parts." Paul Humphreys in "How Properties Emerge" (1997) discusses a "fusion" operation whereby property instances of components at one level "combine" to yield a property instance of a whole at a distinct higher level. While Humphreys speaks of fusion of property instances at lower levels, he rejects the idea that emergents supervene on property instances at the lower levels.

The part/whole aspects of the emergentist doctrine are clearly related to the conception of the world as dividing into distinct levels. I have no quarrel with the claim that the world divides into levels, though I think most philosophers are too simplistic in their characterization of the hierarchy.[6] However, I do disagree with the view that the part/whole aspects stressed in tenets 1 and 2 are essential for the characterization of all types of emergence and emergent properties. There are many cases of what I take to be genuine emergence for which one would be hard-pressed to find part/whole relationships playing any role whatsoever. I also think that many of these cases of emergence do not involve different levels of organization. It will come as no surprise, perhaps, that these examples arise when one considers asymptotic limits between different theories.

The third and fourth tenets of emergentism refer to the unpredictability and the irreducibility/unexplainability of genuinely emergent properties. Much of the contemporary literature is devoted to explicating these features of emergentism.[7] By considering various examples of properties or structures that emerge in asymptotic limiting situations, we will see in chapter 8 that these tenets of emergentism also require emendation. In the course of this discussion we will also come to see that the close connections between reduction and explanation must be severed. In many instances, it is possible to explain the presence of some emergent property or structure in terms of the base or underlying theory; yet the property or structure remains irreducible.

Furthermore, attention to asymptotic limits will reveal that there are borderlands between various theories in which these structures and properties play *essential explanatory roles*. Recognition of this fact may take us some distance toward interpreting (or re-interpreting) the fifth tenet of emergentism noted by Kim. This is the claim that emergent properties must possess novel causal powers—powers that in some sense are not reducible to the causal powers of their basal constituents. Instead of speaking of novel causal efficacy, I think it makes more sense to talk of playing novel explanatory roles. Reference to these emergent structures is essential for understanding the phenomenon of interest; and, furthermore, no explanation can be provided by appealing only to properties of the "basal constituents," if there even are such things in these cases.

One example I will discuss in detail is that of the rainbow. There are certain structural features of this everyday phenomenon that I believe must be treated as emergent. Such features are not reducible to the wave theory of light. The full explanation of what we observe in the rainbow *cannot* be given without reference to structures that exist only in an asymptotic domain between the wave and ray theories of light in which the wavelength of the light $\lambda \to 0$. This is a singular limiting domain, and only attention to the details of this asymptotic domain will allow for a proper understanding of the emergent structures.

Furthermore, we will see that the emergent structures of interest—those that dominate the observable phenomena—typically satisfy the requirements of universality. In other words, the emergent structures are by and large detail-

---

[6] A notable exception is William Wimsatt. See his article Wimsatt (1994) for an extended discussion of the idea of levels of organization.

[7] Kim (1999, pp. 2–3) argues, persuasively I think, that so-called nonreductive physicalism discussed in the philosophy of mind/psychology literature is the modern brand of emergentism. This literature focuses on questions of reduction and explanation.

independent and are such that many distinct systems—distinct in terms of their fundamental details—will have the same emergent features.

2.5 Conclusion

This chapter has suggested that our philosophical views about explanation, reduction, and emergence would better represent what actually takes place in the sciences were they to recognize the importance of asymptotic reasoning. The explanation of universal phenomena, as much of the rest of the book will try to argue, requires principled means for the elimination of irrelevant details. Asymptotic methods developed by physicists and mathematicans provide just such means. Furthermore, much of interest concerning intertheoretic relations can be understood only by looking at the asymptotic limiting domains between theory pairs. This is something upon which virtually no philosophical account of theory reduction focuses. Chapters 6 and 7 discuss in detail the importance of this view of intertheoretic relations. Finally, chapter 8 will expand on the claims made here to the effect that genuinely emergent properties can be found in the singular asymptotic domains that exist between certain pairs of theories.

3

Philosophical Theories of Explanation

Section 2.2 focused on explicating in a fairly intuitive way the sense of universality that is often of interest in many contexts. In particular, universal behavior often seems surprising and therefore cries out for explanation. In this chapter I want to argue that the (asymptotic) methods developed over time by physicists and applied mathematicians to explain instances of universality constitute a type of explanation that has been virtually overlooked in the extant philosophical literature on scientific explanation. These methods do, indeed, represent a distinct form of explanatory reasoning.

3.1 Different Why-Questions

To begin, it is useful to make explicit a distinction that played an important, if somewhat implicit, role in the discussion of the behavior of struts in section 2.1. In asking for an explanation of a given phenomenon such as the buckling of a strut, one must be careful to distinguish between two why-questions that might be being asked. In an earlier article, Batterman (1992), I called these type (i) and type (ii) why-questions. A type (i) why-question asks for an explanation of why a given instance of a pattern obtained. A type (ii) why-question asks why, in general, patterns of a given type can be expected to obtain. Thus, a request to explain an instance of universality is a request to provide an answer to a type (ii) why-question.

I could continue to discuss this in terms of explaining the behavior of struts. However, another example will make more clear the role played by limits and asymptotic reasoning.

So, consider the following game, sometimes called "the chaos game."[1] To play the game, one marks off three vertices of a triangle on a piece of paper.

---

[1] This example is discussed in quite some detail in Batterman (1992).

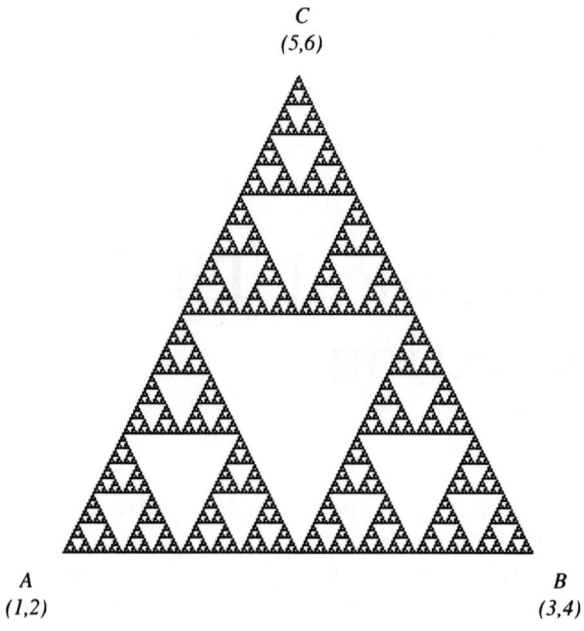

Figure 3.1: Sierpinski triangle

Label the first "A" and assign it the numbers 1 and 2, label the second "B" and assign it the numbers 3 and 4, and label the third "C" and assign it the numbers 5 and 6. Choose one point (actually it doesn't matter whether the point is even in the "triangle space") as a starting point and begin rolling a six-sided die. Suppose we chose point A as our starting point and the first roll landed on the number 4. This number is assigned to point B. The rules say to move halfway from the starting point toward point B and make a mark. This is now our new "starting point." On the next roll we do the same thing. Suppose the die landed on 5. We move halfway toward point C and make a mark there. Continuing to play, we find after many iterations the pattern shown in figure 3.1 called the "Sierpinski triangle." The Sierpinski triangle is a fractal—a figure with a noninteger dimension approximately equal to 1.58.

The question is: Why do we get this pattern when we play the chaos game? Of course, this is an ambiguous question. Do we mean to be asking a type (i) question or a type (ii) question? If the former, then we are asking for an explanation of why the individual dots on the page appear where they do. If the latter, we are asking for an account of why, whenever we play the chaos game, we get patterns that look like this—patterns, that is, characterized by a fractal dimension of 1.58.

Suppose we meant to be asking the type (i) question. What counts as a decent answer? That is, what is the explanation for the appearance of the pattern on the page? Well, we started at point A and rolled a long sequence of numbers from the set {1, 2, 3, 4, 5, 6}. This corresponds to a long sequence of A's, B's, and C's. In particular, the sequence was: B, C, C, B, A, B, B, B, C, A, B, A, A, B, C, ... After each roll we followed the rules of the game. These facts apparently explain why each dot on the page appeared where it did. In the end, we simply *see* that we get a pattern like that in figure 3.1. We might further fill in this account by specifying the initial conditions of the die, the equation that describes its motion, and so on. Were we to be able to do all of this, we will have provided one of Railton's complete ideal explanatory texts.[2] We will have given a causal-mechanical explanation of the appearance of the pattern on the page.

On the other hand, had we meant to ask the type (ii) question, the answer just given would be of no help. This is true for the same reason that the detailed story about the buckling behavior of a particular strut in no way can explain why struts, in general, buckle the way that they do. The causal-mechanical details of a given system cannot provide an explanation of universal behavior.

Before discussing what does answer the type (ii) question about the pattern displayed in figure 3.1, it is best to back-track a bit and discuss various features of rival philosophical accounts of scientific explanation.

3.2 Hempelian Explanation and Its Successors

Contemporary views about explanation and understanding are all the results of reactions one way or another to Carl Hempel's influential account of explanation. In fact, as is well known, most discussions begin with some brief explication of the Hempelian Deductive-Nomological (D-N) model. I am not going to try to do justice to all of the niceties of Hempel's own views or to the responses and theories of his successors. Nevertheless, let me briefly offer what I hope is not too much of a caricature of these different accounts.

Hempelian Models

Hempel held that explanations are essentially arguments or derivations. He was motivated, in part, by a desire to have a model that would be appropriate to different scientific theories. With the advent of the quantum theory, for example, one apparently needs an account of scientific explanation that doesn't depend on an appeal to *mechanism* as traditionally understood.[3] We need, that is, an account of explanation that will apply to phenomena for which (as far as we know) there are no mechanisms—no gears or pulleys—to which we can point. Thus, Hempel's aim was to develop a model that, quite generically, required an appeal to generalities or laws.

Hempel conceived of explanation primarily on the model of solving initial value problems for ordinary differential equations. For example, we explain why

---

[2] Recall the discussion of section 2.1.
[3] By "mechanism" here, I mean to refer to action-by-contact "deterministic causes" such as gears, pulleys, incline planes, and so on.

the planets obey Kepler's laws by deductively deriving those laws or approximations to them (there are interesting issues here) from Newton's laws of motion. D-N explanations such as this have the following schematic form:

$$\frac{\begin{array}{ccc} L_1 & \ldots & L_n \\ C_1 & \ldots & C_m \end{array}}{E}$$

On this view, explanation involves the deductive subsumption of the explanandum E under the (dynamical) laws L_i of the appropriate theory together with certain contingent matters of fact C_j—typically initial conditions.

In essence, the idea is that one has an explanation of some fact or phenomenon when one can derive that fact from the appropriate laws and initial conditions. Thus, in the case of laws represented by (ordinary) differential equations, we have an explanation of some fact, say, why the system is in state S_f at time t_f, if we can solve the equations of motion given the system's state at some other time t_i.

Hempel thought that phenomena governed by deterministic laws would have explanations that conformed to the D-N schema. In particular, the laws L_i are to be "strictly universal" in form. Schematically, Hempel took these to have the logical form of universal conditionals:

$$(\forall x)(Fx \rightarrow Gx).$$

This contrasts with another type of law found in theories that govern indeterministic (or, perhaps, incompletely characterized)[4] phenomena. These laws, according to Hempel (1965, pp. 378–380), are of probabilistic or statistical form. On Hempel's view these laws have the form of conditional probability statements: $P(F|G) = r$—the probability of a's being F given that G equals r.[5]

On Hempel's view, explanations of phenomena governed by statistical or probabilistic laws conform to what he called the "Inductive-Statistical" (I-S) model:

$$\frac{\begin{array}{cc} Pr(G|F) & = & r \\ Fa & \end{array}}{Ga}.$$

In this model the inference is not deductive as the double line indicates. Furthermore, the degree of support of the explanandum Ga, given the explanans, can be at most r.

[4]Possibly, classical statistical mechanics is an example of such a theory. There are many issues of deep consequence in how to characterize this theory. See Sklar (1993) for details.
[5]Of course, Hempel was well aware that the distinction between lawlike and accidental generalizations cannot be characterized by appeal to logical form alone. Nevertheless, he did hold that there was an important distinction between laws in deterministic theories and those in statistical or probabilistic theories. Furthermore, at least part of the distinction between these *types* of laws could be captured in terms of their logical form.

Hempel further required that for an argument of the I-S form to be explanatory, the value of r needs to be high. "[A]n argument of this kind will count as explanatory only if the number r is fairly close to one" (1965, p. 390). This requirement leads to one serious problem (among several) with I-S explanations: There is no nonarbitrary value in the real interval between zero and one (other than 1/2, perhaps). What could count as the cutoff below which the argument fails to explain and above which it would be explanatory?

Even worse, it seems that some happenings are objectively improbable. An α-decay of a given U^{238} nucleus in a specific one-hour time span (and in the absence of external radiation) is exceedingly improbable. (The half-life of U^{238} is on the order of 10^9 years.) Nevertheless, if we witness such a decay and would like to explain its occurrence, it seems that the best we can say is that the quantum theory tells us that the probability of that event is exceedingly small. Surely, being told that there is such a small chance, and that there are no hidden variables that would allow us to determine with certainty whether and when such a decay will occur, is crucial explanatory information. If the event is objectively improbable, then it is unlikely, although not impossible, for it to occur. And we shouldn't expect our theory of explanation to tell us that its occurrence was likely. This is one of the motivations behind Railton's (1981) Deductive-Nomological-Probabilistic (D-N-P) model of explanation. Let me now briefly discuss some aspects of this model before returning to the discussion of the chaos game.

Railton's D-N-P Model

Railton's D-N-P model is motivated in part by a desire to address the question of how one can explain events or happenings that are objectively improbable. The model is less of a departure from the D-N model than is Hempel's own I-S model in that it retains the view that explanatory arguments are deductive in form. However, it pays a price for this. Unlike the I-S model, the D-N-P model is restricted in scope to those phenomena that are objectively indeterministic or probabilistic. Thus, for Railton, the D-N model forms the backbone of the ideal explanatory texts that are required for full explanations of deterministic phenomena, while the D-N-P model plays this role if the phenomenon is indeterministic. This fact is reflected in the form probabilistic laws are taken to have according to the D-N-P model.

Statistical or probabilistic laws have a completely different logical form from Hempel's statements of conditional probability, $P(F|G) = r$. For Railton a genuinely probabilistic law is, just like a deterministic law, a statement having the form of a universal conditional:[6]

$$(\forall x)(\forall t)(G(x,t) \rightarrow Pr(F(x,t+\epsilon)) = r), \text{ for } \epsilon \geq 0. \tag{3.1}$$

D-N-P explanations now take the following form. First, there must be derivation from the underlying theory of a probabilistic law of the form (3.1). Given this,

---

[6] "$Pr(F(x,t)) = r$" says "the probability that x is F at t is r."

the rest of the argument fits the following schema:

$$(\forall x)(\forall t)(G(x,t) \rightarrow Pr(F(x,t+\epsilon)) = r), \text{ for } \epsilon \geq 0$$
$$\underline{G(a,t_0)\hspace{6cm}}$$
$$Pr(F(a,t_0+\epsilon)) = r \text{ [and } F(a,t_0+\epsilon)]$$

The addendum in the last line is the explanandum; for instance, that the particular U^{238} nucleus decayed at a certain time. Strictly speaking, it is not part of the argument, nor does it follow deductively from the premises. Nevertheless its inclusion is necessary if we want to string together D-N-P and (possibly) D-N arguments so as to provide the skeleton of ideal explanatory texts for a given phenomenon.

Railton departs from the strictly Hempelian view that explanations are "purely arguments." Deductive arguments of the D-N and D-N-P forms often play key roles in explanation on this view, but they are not (or usually not) by themselves sufficient for explanation. Nor are arguments of these forms always necessary. Sometimes all that is required for an answer to a why-question to be considered explanatory is that the answer remove some degree of uncertainty about the form of the explanatory text. The key for Railton, and this is why his view fits the causal-mechanical mold, is to be able to provide (arbitrary parts of) the appropriate ideal explanatory text. These texts, as we've seen, are detailed causal-mechanical accounts of the workings of the mechanisms leading to the occurrence of the explanandum phenomenon. This means that part of what it is to give the causal-mechanical details is the providing of D-N and D-N-P arguments—depending on whether the mechanisms are deterministic or probabilistic in nature.

We've seen what the ideal explanatory text for the presence of the given instance of the dot pattern on the paper after one round of the chaos game will look like. It will contain D-N argument patterns as its backbone, since the rolling of a die isn't usually taken to be an indeterministic or irreducibly probabilistic mechanism. It seems perfectly reasonable to me to hold that such a text provides much, if not most, of the explanatory information about the occurrence of that one instance. It tells us all of the gory details about why each dot appears where it does.

However, we have also seen that such a text cannot answer the type (ii) why-question. It gives us no reason to believe that the next time we play the game we will get similar behavior—that is, a pattern with the same fractal dimension. So what does explain it?

Note that it is not the case that every sequence of rolls of a fair die will yield triangle patterns of the appropriate sort—patterns that we must in some sense expect to occur. After all, a perfectly possible outcome might be the sequence: A, B, A, B, A, B, A, B, A, B, A, B ... (or some other intuitively "nonrandom" sequence). We expect sequences like these to be unlikely outcomes of the repeated rolling of a fair die and it is possible to make precise this assessment of likelihood or probability. Sequences of rolls are sequences of results that are independent and identically distributed: Each roll is probabilistically independent of any other, and each outcome (each number on the die and, hence, each letter A, B, and C) has the same probability of obtaining (respectively, 1/6 for the number and 1/3 for the letter).

For sequences having these properties, one can appeal to various limit theorems of probability theory. For example, a version of the strong law of large numbers holds. One version says that we can expect (*with probability one*) that the relative frequency of A's in an infinite sequence of rolls will equal 1/3—the probability of A on a single roll:[7]

$$Pr(\lim_{n \to \infty} \sum_n \frac{A_n}{n} = 1/3) = 1.$$

By appealing to limit theorems like this, it is possible to argue that patterns like that seen in figure 3.1 are *with probability one* the expected outcomes of playing the chaos game. It looks like we can give an I-S-like explanation for the generic appearance of the triangle pattern by *demonstrating that such patterns are highly probable—in fact, they have unit probability—in an ensemble of sequences of the kind described*.[8]

But the causal-mechanical theorist, particularly one of Railton's stripe, finds herself in a bind here. She cannot appeal to this probabilistic fact to answer the type (ii) why-question. After all, given the deterministic nature of the mechanisms operating in the game, the appropriate schema is the D-N schema. But, if the term "probability" can appear at all in such a schema, it must be used in an expression synonymous with certainty (probability one) or impossibility (probability zero). However, these senses are not measure-theoretic. It is possible, even though the measure-theoretic probability is zero, to have a sequence that fails to yield the triangle pattern. There are infinitely many of them, in fact.

The appeal to limit theorems of this ilk is not restricted to "artificial" or "mathematical" examples like the chaos game.[9] Many instances of I-S arguments utilizing a probability one result can be found in physics. In statistical mechanics and ergodic theory, the aim of theoretical investigation is often to try to justify the appeal to such limit theorems as a result of instabilities in the dynamical evolutions of the various systems. The probability one results required for explanatory arguments of this type block the objection raised earlier about the arbitrariness of the degree of inductive support conferred upon the explanandum by the explanans that should be required, in Hempel's I-S model, for genuine explanation. If the explanandum has probability one, then as far as

---

[7] "A_n" is the random variable taking the value 1, if the nth roll results in A (i.e., in the die showing a 1 or a 2) and the value 0, otherwise.

[8] As I will argue, this isn't the whole story. We also must require an account of why the probability one "law" holds.

[9] Though, as I've described it here, this is not strictly a mathematical example. It requires at the very least, some connection with the dynamics of the die rolling mechanism. See Batterman (1992) for a discussion of more "physical" examples.

statistical explanation goes, there can be no question of having a higher degree of support.

I think that something very much like this partial explanatory strategy functions in many situations in physics. The appeal to (physically relevant) limit theorems is essential. What justifies these appeals, and the limiting relations themselves, are often deep asymptotic properties of the behaviors of systems composed of many components. Indeed, what is most important is an account, grounded in fundamental theory, that tells us why the probability one claim holds. This account will involve asymptotic analysis.[10]

Before continuing the discussion of these issues, however, I need to address a second fundamental development in philosophical views of explanation, one that also clearly has its roots in Hempel's D-N model. This is the recognition of the explanatory virtues of unification.

Unification Theories

Causal-mechanical explanations have been characterized as "local" in the following sense:

> [T]hey show us how particular occurrences come about; they explain particular phenomena in terms of collections of particular causal processes and interactions—or, perhaps, in terms of noncausal mechanisms, if there are such things. (Salmon, 1989, p. 184)

However, there is a view emerging out of the Hempelian D-N tradition defended by Michael Friedman (1974) and Philip Kitcher (1989) (among others) that characterizes explanation "globally," taking it essentially to be a species of unification.

Both views—the causal-mechanical and the unification—hold, I believe, that part of the aim of a theory of explanation is to explicate the concept of scientific understanding.[11] On the causal-mechanical conception, scientific understanding is a local affair. As we've seen, one gains understanding of a given part of the world (considered initially as an opaque black box) by elucidating the mechanisms that operate within that black box.

The goals of the unification theory vis-à-vis scientific understanding are quite different. On this account, roughly speaking, understanding is achieved by reducing the number of independent assumptions one must make when explaining the way things are in the world. Unification is a global concept: "Explanations

---

[10] As one can see, my main complaint with Railton's account of explanation, and with causal-mechanical accounts in general, is that they miss the role played by asymptotic reasoning. One can criticize Railton for his strict elimination of probabilities from explanations of deterministic phenomena as I have done in Batterman (1992). But, the point here is more fundamental. The attention to all of the causal details required for an ideal explanatory text simply won't allow for an appeal to the limit theorems that are applicable, precisely because those details can and must be ignored.

[11] An alternative view is that of Schurz and Lambert (1994), who offer a unification-based account of understanding that does not depend upon a model of explanation.

serve to organize and systematize our knowledge in the most efficient and coherent possible fashion. Understanding ... involves having a world-picture—a *scientific Weltanschauung*—and seeing how various aspects of the world and our experience of it fit into that picture" (Salmon, 1989, p. 182).

From this superficial description, it is already clear that the two accounts present two quite different senses of explanation and understanding. Nevertheless, Salmon argues that

> [t]hese two ways of regarding explanation are *not incompatible* with one another; each one offers a reasonable way of construing explanations. Indeed, they may be taken as representing two different, but compatible, aspects of scientific explanation. Scientific understanding is, after all, a complex matter; there is every reason to suppose that it has various different facets. (1989, p. 183)

It would be nice if these two senses of explanation and understanding mapped nicely onto the two types of why-questions highlighted. Unfortunately, while I think one can profitably understand the causal-mechanical conception as providing appropriate responses to type (i) but not type (ii) questions, it is *not* straightforward (and perhaps not even possible) to see the unification models as always providing appropriate answers to type (ii) questions about universal behavior.

However, the contrast between the two explanatory models has sometimes been pitched in just this way (though not in so many words). Consider an example discussed by Kitcher in his article "Explanatory Unification and the Causal Structure of the World" (1989). The example concerns a party trick in which a cord is wrapped around a pair of scissors in such a way that it appears to be knotted. Yet if the victim of the trick makes an nonobvious twist at the start of trying to untie it, the cord is easily removed. One would like an account of why the victim typically cannot disentangle the cord from the scissors. Kitcher says,

> [W]e could, of course, provide the causal details, showing how the actions actually performed lead to ever more tangled configurations. But this appears to omit what should be central to the explanation, namely the fact that the topological features of the situation allow for disentanglings that satisfy a specifiable condition, so that sequences of actions which do not satisfy that condition are doomed to failure. (1989, p. 426)

This is meant to suggest that "even when causal histories are available, they may not be what the explanation requires" (Kitcher, 1989, p. 426). I take it that the spirit of this objection is exactly the same as my criticisms of possible causal-mechanical attempts to explain universal behavior.

But can the unification theories answer type (ii) why-questions? Here we need at least to sketch a version of the unification theory. I will focus on Kitcher's version since it is the most developed and has received the most attention in the literature.

Both Kitcher and Friedman hold that arguments which unify "phenomena" provide understanding and count as explanatory in virtue of their unifying ability. Kitcher's version of explanatory unification is based on the idea that the same relatively small set of patterns of derivation is used repeatedly to derive many different conclusions—explananda.

> Understanding the phenomena is not simply a matter of reducing the "fundamental incomprehensibilities" but of seeing connections, common patterns, in what initially appeared to be different situations.... *Science advances our understanding of nature by showing us how to derive descriptions of many phenomena, using the same patterns of derivation again and again, and, in demonstrating this, it teaches us how to reduce the number of types of facts we have to accept as ultimate (or brute).* (Kitcher, 1989, p. 432. Emphasis in original.)

On this view, a derivation counts as an explanation if it is a member of a distinguished set of derivations $E(K)$, called the "explanatory store over K." Here K is, roughly, the set of statements accepted by the scientific community, and $E(K)$ is the set of derivations that best "systematizes" K. The criterion, of course, for systematization is unification (Kitcher, 1989, p. 431). The challenge for the unification theorist is to make the notion sufficiently precise as to be actually useful *and* enlightening.

For present purposes, one need not go into the details of Kitcher's proposal. Nevertheless, it is worth pointing out a few key features. First, whether a given derivation counts as an explanation depends, crucially, on its being a member of the set $E(K)$ that best unifies our "scientific knowledge" K. That is, a given D-N argument, say, will not be explanatory unless it is a member of $E(K)$. This reflects the *global* nature of explanatory unification. Second, at base there are two competing criteria for unification. The conception of unification endorsed by Kitcher involves the idea that the explanatory store $E(K)$ "is a set of derivations that makes the best tradeoff between minimizing the number of patterns of derivation employed and maximizing the number of conclusions generated" (1989, p. 432). Third, it is instructive to characterize the unification proposal as a kind of "explanatory internalism." This terminology comes from Jaegwon Kim's discussion of explanation in his article "Explanatory Knowledge and Metaphysical Dependence" (1994). Let me elaborate on this aspect of explanatory unification.

Kim contrasts explanatory internalism with "explanatory externalism" or "explanatory realism." An internalist model of explanation treats explanation as

> an activity *internal* to an epistemic corpus: whether or not something is an explanation—a good, "true" or "correct" explanation—depends on factors internal to a body of knowledge, not what goes on in the world—except, of course, for the truth of the statements comprising the explanans.... A realist about explanation believes

that some objective relation between the [explanandum event and the event(s) invoked as its explanation] underlies, or grounds, the explanatory relation between their descriptions. (Kim, 1994, p. 57)

As Kim sees it, Kitcher's account is internalist because whether a given derivation counts as an explanation depends on its inclusion in $E(K)$, where, as we have seen, $E(K)$ is determined solely by "factors internal to the epistemic system, such as the number of argument patterns required to generate the given class of arguments, the 'stringency' of the patterns, etc., not on any objective relations holding for events or phenomena involved in the putative explanations" (Kim, 1994, p. 64).

Now how might a unification theorist try to answer the type (ii) why-question about the existence of the triangle pattern in figure 3.1? It seems that what the theory requires is that she add the argument pattern (I-S) to the explanatory store $E(K)$, with the further stipulation that the probability in the law schema, r, is unity. The justification for doing this would be that there are many different probability one type claims that get used in diverse areas involving deterministic theories, so that adding this one pattern will better unify our scientific knowledge K. Thus, the unification theorist explains the existence of the triangle pattern by pointing out that *there is an argument schema in $E(K)$ that tells us to expect the pattern with probability one.*

However, in a very important sense, I think this misses the point. The problem is that there is more to understanding why the pattern emerges than is provided by noting that it can be expected with probability one. One would like to know why the probability one result—the statistical law—itself holds. Answering this would provide the ultimate explanation for the universal behavior. Recall that Railton's model has the virtue of demanding that the law appearing in the D-N-P schema be derived from a fundamental probabilistic theory such as quantum mechanics. I think that something similar must be required here as well. We need to understand, by appeal to fundamental theory, why the probability one claim holds. I will consider the nature of various "arguments" that lead in different cases to conclusions of that form later. They all have the following feature in common. Asymptotic analyses show that many, if not most, of the details required for characterizing the types of systems in question are irrelevant. But in each case that reasoning may be very different and the form of the limiting "laws" may also be quite different. (I will have much more to say about this.) I do not think that the purely internalist account of explanatory unification can easily respect these differences. That is to say, I don't think the unification theorist can simply add to $E(K)$ some schema that says "use asymptotic analysis."

It is worth returning briefly to Kitcher's example of the party trick involving the "knotted scissors." I think that his own discussion of what is explanatory in this situation is suggestive, but also that it doesn't really fit the unification scheme. Recall that what we would ultimately like to explain is the fact that the victim of the trick fails to untie the "knot." Kitcher says that what is central to the explanation is "the fact that topological features of the situation allow for disentanglings that satisfy a specifiable condition, so that sequences of actions which do not satisfy that condition are doomed to failure" (1989, p. 426). For Kitcher, the explanation involves identifying the particular phenomenon (the failure to disentangle the cord in a given case) as "belonging to a class—typically defined in terms of some language that does not occur in the initial posing of the question.... The failure to untangle the telephone cord is explained by using topological notions to characterize the initial configuration of the wire" (1989, p.427). Thus, "[w]e explain a victim's frustration with the telephone cord by identifying the topological features of the 'knot,' and noting that only certain kinds of actions will produce the desired result" (1989, p.427).

If we add to the explanatory store $E(K)$ a generalization (an axiom?) to the effect that all cord-plus-scissors systems having topological property P are such that only twists of kind Q result in untanglings, then it is easy to "explain" why the particular victim fails to untangle the "knot."[12] What may make this seem like some sort of advance is, as Kitcher stresses, the fact that we have appealed to terms from a "language" (the language of topology) different than that used to pose the initial why-question. But is the generalization itself the conclusion of some further derivation that is an instance of a pattern in $E(K)$? It seems that the only instance of an argument pattern in $E(K)$ that would have this generalization as a conclusion is one in which the generalization itself would have to appear as a premise. Hence, it looks like the unification model is incapable of explaining the generalization.

Suppose, though, that we have analyzed the topological structure of an abstract space in which one can represent all possible manipulations of the cord and scissors. Suppose, further, that our analysis has yielded some conclusions about the *stability* of the "knot" under similar but distinct manipulations. Then we may be able to understand why the majority of such manipulations fail to untangle the cord. That is, we may understand why the generalization we added to $E(K)$ is true. Such an analysis gives us an idea of what detailed features of the "knot" are *irrelevant* for the characterization of the dominant behavior of interest. In a certain sense it allows us to classify all (or at least most) of the causal details Kitcher mentions as explanatory "noise" which can be ignored. Understanding of universal behavior is achieved, at least in part, when one has a topological analysis of an appropriate abstract space providing principled reasons for ignoring certain details. It is an appeal to stabilities of a certain kind in such a space that ultimately grounds the feeling that the victim's failure to untie the "knot" is explained by derivation from the generalization that makes reference to the topological property P. However, I do not believe that such an analysis can easily be construed itself as a derivation of the generalization from other statements in K.

---

[12]This proposal fits with an argument by Todd Jones in which he argues that sometimes (in cases where multiple realizability of upper-level properties or laws blocks "reduction"), "the explanatory store must either include such high-level generalizations as axioms, or be capable of generating them with other high-level patterns" (1995, p. 27). I argue in Batterman (2000) that instances of multiple realizability are profitably understood as instances of universality. This will be discussed in more detail in chapter 5.

Of course, all of this is still very vague. The case we are imagining is too underdescribed. Here I hope to have shown only that there are serious concerns about how the purely internalist model of explanatory unification can provide explanations for patterns of a certain sort. It remains to be seen whether the vague suggestions about appeals to asymptotic analyses and to stabilities of mathematical structures in appropriate abstract spaces provide the beginning of a theory of understanding—one that will account for the worries expressed by Fisher in the passage quoted in section 2.1. We need to examine more specific examples and see if something like this suggestion is correct.

3.3 Conclusion

I have considered how various philosophical models of explanation fare when it comes to the explanation of universality. I have argued that neither the causal-mechanical models nor the unification models manage to do a very good job. As we have seen, the local causal-mechanical models are worse off than the global approaches that appeal to unification. However, both models virtually ignore the sort of asymptotic reasoning that, as we will see, is necessary for an explanation of universal behavior.

In the next chapter I will discuss the explanation of the universality of critical phenomena provided by renormalization group arguments in physics.[13] This is an example of one of the best understood and most systematic explanations of universality. However, there are many other examples in which reasoning similar to the renormalization group analysis plays the essential role in explanation. I will examine some other, superficially very different, examples and will try to outline the basic features of "asymptotic explanation."

---

[13] I have discussed this example in different contexts in two articles, Batterman (1998, 2000).

4

Asymptotic Explanation

In this chapter I will present several examples of explanation in which asymptotic reasoning plays an essential role. After discussing the renormalization group explanation of the universality of critical phenomena, I will, in section 4.2 highlight those features of asymptotic explanation that are likely to generalize. Then in section 4.3 I present two more examples involving explanatory asymptotic reasoning in the long-time regime. Finally, in section 4.4 I will sum up the discussion of asymptotic explanation and discuss the important role played by stability considerations. We can see from this discussion that asymptotic explanation does, in fact, represent a distinct form of explanation largely missed by current philosophical conceptions.

4.1 The Renormalization Group (RG)

I remarked earlier that the most familiar examples of the use of the term "universality" appear in the physics of critical phenomena. It is worth understanding how physicists have explained the remarkable fact that systems of very different microscopic constitution exhibit identical behavior under certain circumstances. Physicists and mathematicians have developed a fairly systematic strategy through which this explanation can be effected. It is called the renormalization group.

While the details of this explanatory strategy can seem quite daunting, the basic idea is very simple. In this and the next section I will try to emphasize exactly what is crucial and "new" (at least from the point of view of philosophical models of explanation) about the renormalization group analysis, and "asymptotic explanation" in general.

To see how the explanation is supposed to go, I must first introduce a few details about critical phenomena in systems such as fluids, gases, and magnets. Experimentally it is found that certain dimensionless numbers, so-called critical exponents, apparently characterize the (virtually) identical behavior of systems as diverse as fluids and magnets near their respective critical points. The critical point for a fluid in a container, for instance, is a point below which it is impossible to pass, by increasing pressure at fixed temperature, from the gaseous phase to the liquid phase without going through a state in which both gas and liquid are simultaneously present in the container, and above which such a transition from gas to liquid is possible without passing through a regime in which both gas and liquid are simultaneously present. That is, for fixed $T > T_c$ (the critical temperature), one no longer observes a vapor phase followed by a liquid phase as the pressure is increased. Instead, there is a unique "fluid" phase. Figure 4.1 illustrates this. In the figure the lines represent values of P and T for which two phases are simultaneously present. The point A is a "triple point" at which all three phases can coexist.

The critical exponent describes the behavior of the density of the fluid as a function of temperature near the critical point. It describes a universal property in that the same behavior is observed almost without regard for the chemical constitution of the fluid. More remarkable, still, is the fact that the *same* critical exponent apparently describes the behavior of magnets as they undergo a transition from the ferromagnetic state with positive net magnetization below the critical point to the paramagnetic phase with zero magnetization above the critical point. Surely, one would like an explanation of this universal behavior. Modern statistical physics (RG arguments) provides such an account.

We see here, explicitly, the two features of universality highlighted in section 2.2. First, the *details* of the microstructure of a given fluid are largely irrelevant for describing the behavior of the particular system of interest. Second, many different systems (other fluids and even magnets) with distinct microstructures exhibit identical behavior characterized by the same critical exponent. As we will see, the RG analysis of critical phenomena will explain why the different systems exhibit the same behavior (the second feature) by utilizing the fact that at the critical point, the details of the microstructure of any one individual system are virtually irrelevant for *its* behavior at that point (the first feature).

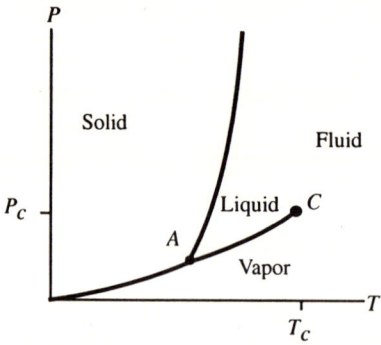

Figure 4.1: Temperature-pressure phase diagram for a "fluid"

Asymptotic Explanation

The critical behavior of a fluid is best characterized (at least at the simplest level) in terms of what is called the "order parameter." In this case the order parameter is the difference in densities of the vapor and liquid present in the container of "fluid": $\Psi = \rho_{liq} - \rho_{vap}$. Below the critical temperature, T_c, Ψ is nonzero, indicating the simultaneous presence of the two phases of the fluid; while above T_c the order parameter Ψ vanishes. Figure 4.2 illustrates the key points. In particular, the vertical lines tie together values for coexisting liquid and vapor densities, (ρ_{liq}) and (ρ_{vap}) respectively.[1]

One universal aspect of the behavior of fluids near the critical point is the shape of the coexistence curve drawn in figure 4.2. The key question concerns the shape of this curve near the critical temperature T_c. To begin to answer this, one considers the "reduced" temperature

$$t = \frac{T - T_c}{T_c},$$

which gives the difference in temperature from the critical temperature in dimensionless units. One makes the assumption (well supported by experimental data) that as T approaches T_c from below, the order parameter Ψ vanishes as some power β of $|t|$:

$$\Psi = \rho_{liq} - \rho_{vap} \sim |t|^\beta.$$

If the coexistence curve were parabolic near T_c, then β would be equal to $1/2$. Experiments have shown, however, that the exponent β is not equal to $1/2$, but rather has a value near 0.33 that is also, apparently, not a simple rational fraction. Furthermore, the *same* number appears to characterize the coexistence curves for different fluids and even, as noted earlier, coexistence curves for magnets where the order parameter is the net magnetization, M. The magnetization is positive below the critical temperature T_c, the system being in a ferromagnetic phase and is zero above T_c with the system in a paramagnetic phase:

$$\Psi = M \sim |t|^\beta.$$

We can ask about a given fluid, say F, why near its critical temperature its density varies with temperature the way that we observe it to. We might attempt to answer this starting from fundamental physics by appealing to the Schrödinger equation for the fluid considered to be some sort of collection of molecules interacting according to certain specified force laws. This would be an attempt to fill in part of a causal-mechanical ideal explanatory text. If we were lucky in our attempt to solve the equation we might, in fact, be able to derive the fact that for fluid F, $\Psi_F \sim |t|^{0.33}$. On the other hand, as has been repeatedly noted, such a derivation is *not* an explanation of the universality I am discussing. In no way can it answer the question of why the same exponent describes the critical behavior of fluids F', F'', and so on.

The desired explanation is provided by the RG. To see how this goes, first note that every system, each different fluid, say, is represented by a function—its so-called Hamiltonian. This characterizes the kinds of interactions between

---

[1] Note that the vertical "tie lines" in figure 4.2 are *not* part of the "coexistence curve."

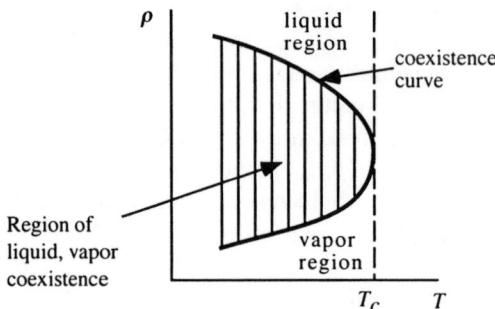

Figure 4.2: The coexistence curve: Density vs. temperature for a Fluid

the system's components (or degrees of freedom), the effects of external fields, and so on. Usually, the different components of a gas or fluid are correlated with one another only very weakly. In other words, they interact significantly only with nearby components and they remain virtually uncorrelated with remote components. However, when a system is near its critical point, the "length" of the range of correlations between the different components increases despite the fact that the interactions remain local. In fact, at the critical point, this so-called correlation length diverges to infinity. The divergence of the correlation length is intimately associated with the system's critical behavior. It means that correlations at every length scale (between near as well as extremely distant components) contribute to the physics of the system as it undergoes a phase transition. This is a highly singular mathematical problem and is, in effect, completely intractable. It is relatively easy to deal with correlations that obtain between *pairs* of particles, but when one must consider correlations between three or more—even more than 10^{23}—particles, the situation is hopeless.

The guiding idea of the renormalization group method is to find a way of turning this singular problem into a tractable, regular problem by transforming the problem into an analysis of the topological structure of an appropriate abstract space—the space of Hamiltonians.[2] One introduces a transformation on this space that changes an initial *physical* Hamiltonian describing a real system into another Hamiltonian in the space. The transformation preserves, to some extent, the *form* of the original Hamiltonian so that when the thermodynamic parameters are properly adjusted (renormalized), the new renormalized Hamiltonian describes a *possible* system exhibiting similar behavior. Most importantly, however, *the entire process effects a reduction in the number of coupled components or degrees of freedom within the correlation length.* Figure 4.3 gives an idea of what is involved here for a spin system modeled on a lattice. In this particular example, the number of degrees of freedom is reduced by a factor of four every time the transformation is iterated.

---

[2] See Pfeuty and Toulouse (1977) for a nice discussion of how the RG works.

Asymptotic Explanation

This kind of analysis is possible because at criticality the divergence of the correlation length means that no characteristic length scale is associated with the system. As a result, the system at criticality exhibits *self-similar* behavior—roughly, it looks the same at different length scales. This is seen in figure 4.3.

Thus, the new renormalized Hamiltonian describes a "system" that presents a more tractable problem and is easier to deal with. By repeated application of this renormalization group transformation, the problem becomes more and more tractable until one hopefully can solve the problem by relatively simple methods. *In effect the renormalization group transformation eliminates degrees of freedom (microscopic details) that are inessential or irrelevant for characterizing the system's behavior at criticality.*

In fact, if the initial Hamiltonian describes a system at criticality, then each renormalized Hamiltonian must also be at criticality. The sequence of Hamiltonians thus generated defines a trajectory in the abstract space that in the

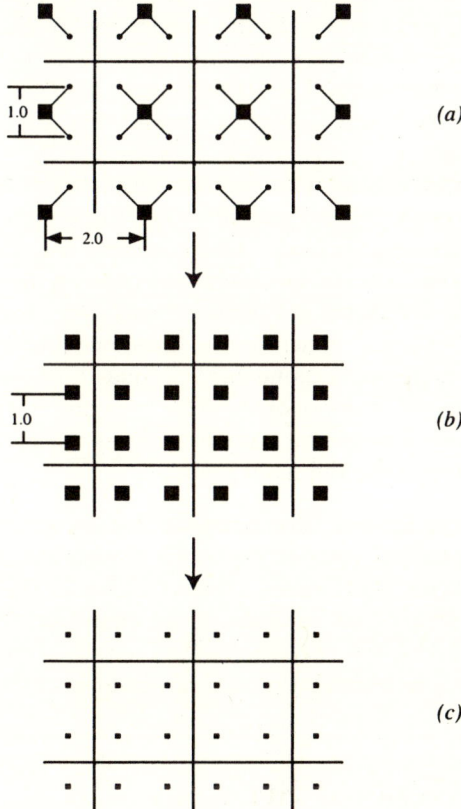

Figure 4.3: The operations of the renormalization group: (a) Block formation (partial integration); (b) Spatial rescaling/contraction; (c) Renormalization of parameters

limit as the number of transformations goes to infinity ends at a fixed point.

The behavior of trajectories in the neighborhood of the fixed point can be determined by an analysis of the stability properties of the fixed point. *This analysis also allows for the calculation of the critical exponents characterizing the critical behavior of the system.* It turns out that *different physical Hamiltonians* can flow to the same fixed point. Thus, their critical behaviors are characterized by the same critical exponents. This is the essence of the explanation for the of critical behavior. Hamiltonians describing different physical systems lie in the basin of attraction of the same renormalization group fixed point. This means that if one were to alter, even quite considerably, some of the basic features of a system (say, from those of fluid F to fluid F'), the resulting system (F') will exhibit the same critical behavior. This *stability* under perturbation is a reflection of the fact that certain details concerning the microconstituents of the systems are individually largely irrelevant for the systems' behaviors at criticality. Instead, their *collective properties* dominate their critical behavior, and these collective properties are characterized by the fixed points of the renormalization group transformation (and the local properties of the flow in the neighborhood of those points). We can, through this analysis, understand both how universality arises and why the diverse systems display identical behavior.

By telling us what (and why) various details are irrelevant for the behavior of interest, this same analysis also identifies those physical properties that *are* relevant for the universal behavior being investigated. For instance, it turns out that the critical exponent can be shown to depend on the spatial dimension of the system and on the symmetry properties of the order parameter. So, for example, systems with one spatial dimension or quasi-one-dimensional systems such as polymers, exhibit different exponents than (quasi-) two-dimensional systems like films. The universal behaviors of these systems are, in turn, different than that exhibited by three-dimensional systems such as fluids. The details here aren't crucial for understanding the main point:

- The RG type analysis *demonstrates that many of the details that distinguish the physical systems from one another are irrelevant* for their universal behavior. At the same time, it allows for the determination of those physical features that *are* relevant for that behavior. Finally it allows for the quantitative determination of the nature of the universal behavior by enabling one to identify the critical exponents.

4.2 The General Strategy

It is possible to extract from this example several general features that are present in many explanations of universality offered by physicists. The explanation in the example is effected by developing reasons, grounded in the fundamental physics of fluids, for ignoring detailed microphysical information about any actual system or systems being investigated. The RG is a method for extracting structures that are, with respect to the behavior of interest, detail independent. Of course, many features remain system or detail dependent. In the previous example, each fluid will have a different critical temperature T_c that depends essentially on its microstructural constitution. Nevertheless, the RG method allows for a construction of a kind of limiting macroscopic phenomenology—an upper level generalization, perhaps—that is not tied to any unique microstructural account. The microscopically distinct systems can be grouped into distinct universality classes depending on the macroscopic phenomenology they exhibit.

Nigel Goldenfeld et al. (1989, p. 368) argue that the principles of RG theory constitute a "general framework for extracting phenomenological relations among macroscopic observables from microscopic models that may not be precisely definable." It is a

> method to study the asymptotic behavior of a system, e.g., in the macroscopic limit, where the scale of observation is much larger than the scale of microscopic description.... It should be stressed that phenomenology is not a second-rate incomplete description of a given system, but the most complete description modulo microscopic unknown factors over which we can never have ultimate detailed knowledge. (Goldenfeld et al., 1989, p. 369)

While this last remark seems to reflect limitations on knowability, the RG type arguments are *not* essentially epistemic in nature. They demonstrate, as I've said, that the explanation of universal behavior involves the elucidation of principled physical reasons for bracketing (or setting aside as "explanatory noise") many of the microscopic details that genuinely distinguish one system from another. In other words, it is a method for extracting just those features of systems, viewed macroscopically, that are stable under perturbation of their microscopic details.

As the example from the last section shows, it is essential that the microscopic details do not themselves greatly influence the phenomenological behavior of the system. For instance, the exact structure of the molecules in a given fluid and the precise nature of the forces characterizing the molecular interactions are not important for the critical behavior of that particular fluid. Instead, in the *asymptotic regime* where many molecules find themselves correlated—that is, where the correlation length diverges—it is the collective or correlated behavior that is salient and dominates the observed phenomenological behavior. In order to "see" the stable, universal phenomenological structures, *one must investigate the appropriate limiting regime*. It is, therefore, an essential feature of the method that it describes the relationship between two asymptotically related descriptions of the system.[3]

In sum, the following features are apparently characteristic of RG type explanations of universal behavior.

---

[3] In many instances, the asymptotic relation between the different "levels of description" will depend on differing length scales. But there is no essential reason that length (or size) must be the asymptotic parameter. Long-time behavior or increasing complexity might, for example, be appropriate parameters for some investigations. See the examples in section 4.3.

- As the universal behavior is in some sense emergent in an asymptotic regime, the explanation involves some kind of asymptotic analysis.
- The universality is the result of the stability under perturbation of the underlying microscopic physical details.
- The stability under perturbation (or "structural stability") is explained by the underlying microscopic physics of the systems being investigated.

4.3 "Intermediate Asymptotics"

We have seen that on Goldenfeld's view, the RG is a method for extracting a macroscopic phenomenology (involving universal behavior) from microscopic models. In so doing, we not only can discover what universal macroscopic regularities hold of the system of interest (if we haven't already observed this empirically),[4] but we also can explain, in terms of the underlying microscopic theory, why those macroscopic generalizations hold. Part of what gets shown is that the macroscopic description is insensitive to the microscopic details. One way of understanding this may be to hold that there is no unique microscopic story underlying the macroscopic phenomenology. Given that, one might as well use the "simplest" microscopic model. Goldenfeld et al. put the point as follows:

> A macroscopic description of a system may be insensitive to microscopic details. Thus there is no unique microscopic picture that is consistent with the macroscopic phenomenology. In fact, there is an equivalence class of microscopic models all of which have the same macroscopic phenomenology. One could say that this fact is epistemologically important, but this is also very useful in practice. If one is solely interested in studying universal, macroscopic properties of a system, one can choose the most convenient or simplest (in some sense) microscopic model. (Goldenfeld et al., 1989, p. 359)

There is clearly a pragmatic element involved here. If what we care about is the macroscopic phenomenology, then we can try to proceed with RG-like asymptotic arguments. Nevertheless, there is more to it than simple pragmatic questions of interest. There are reasons why we so often focus on universal behavior as the subject of investigations.

In studying a phenomenon of physical interest, such as the period of pendulums discussed in section 2.2, we often seek to find relationships between the quantity of interest and the other parameters that govern the behavior of the system. Following Barenblatt (1996), let us call the quantity of interest a and the governing parameters $a_1, \ldots, a_k, b_1, \ldots, b_m$. The distinction between the

---

[4]We have already seen an example of "discovering" the form of the generalization given only background knowledge of the subject. This was the example of dimensional analysis applied to the pendulum's period in section 2.2.

a_i's and the b_j's is that the a_i's have dimensions that are independent of the other parameters, whereas the dimensions of the b_j's are expressible as products of powers of the dimensions of the a_i's.[5] The investigator looks for relationships of the form

$$a = f(a_1, \ldots, a_k, b_1, \ldots, b_m).$$

One way of (partially) defining such a function would be provide a table of values containing the values for the a_i's and the b_j's that are of interest and the corresponding value for a in each case. In the age of modern computers with the ability to run impressive numerical calculations, it is often claimed that this is what is required to answer our questions about the quantity a. Barenblatt puts the point as follows:

> Please, it is often said, formulate the problem whose solution you need and you will obtain without special effort a set of tables of all the numerical values you need with the desired (and, of course, paid for) accuracy. To understand this point properly [that is, to see what's wrong with this point of view] we have to recall that the crucial step in any field of research is to establish what is the minimum amount of information that we actually require about the phenomenon being studied. All else should be put aside in the study. This is precisely the path taken by the natural sciences since Newton. Features that are not repeated from case to case and cannot be experimentally recorded are always deliberately thrown out. Thus, the researcher is actually not so much interested in a table of values of the function f ... as in the principal physical laws that determine the major features of this relationship. These principal laws, the basic features of the phenomena can often be seen and understood when analysing the intermediate asymptotics. (Barenblatt, 1996, pp. 91–92)

This attitude resonates with Fisher's discussion of the understanding afforded by the "modern" theorist (section 2.1). Understanding of universal features, as I have been arguing, often requires more than simply providing a numerical calculation from the fundamental governing equation.

For Barenblatt, the elucidation of the "principal laws" and "basic features" provides the desired understanding. These laws and features are determined by analyzing the intermediate asymptotics of the governing equations. We now need to see what he means by this. We can begin to understand this if we think of an Impressionist painting such as Seurat's pointillist "A Sunday Afternoon on the Island of La Grande Jatte." A characteristic of this painting and of the Impressionists, in general, is the fact that if you are too close, you get lost in

---

[5] We have already seen this in our discussion of the pendulum's period. The parameters are m, l, θ, and g. We have $[m] = M$, $[l] = L$, $[\theta] = T$. These dimensions are all clearly independent of one another. On the other hand, the dimension of g, the gravitational acceleration, is definable in terms of the product of powers of the first three: $[g] = M^0 L^1 T^{-2}$. Thus, the first three parameters would be members of the a_i's and the last is a member of the b_j's.

the details—all you see are the various colored points. But also, as with any painting, if you are too far away, you cannot discern the image either. The "best" view—the view where one can best appreciate the image (or "understand" it)—is to be had at some intermediate distance from the canvas.[6]

Something analogous is involved in "any scientific study."

> Thus, the primary thing in which the investigator is interested is the development of the phenomenon for intermediate times and distances away from the boundaries such that the effects of random initial features or fine details in the spatial structure of the boundaries have disappeared but the system is still far from its final equilibrium state. This is precisely where the underlying laws governing the phenomenon (which are what we are in fact primarily interested in) appear most clearly. Therefore intermediate asymptotics are of primary interest in every scientific study. (Barenblatt, 1996, p. 94)

This last claim is probably a bit too strong. However, I think that, broadly construed, it is true of much of what gets studied in physics and other sciences to which mathematicians have applied their skills. In particular, I think, as should be obvious by now, that any investigation that remotely addresses a question related to understanding universal behavior will involve intermediate asymptotics as understood by Barenblatt. Therefore, it is well worth the effort to examine a few explicit examples of this type of asymptotic reasoning.

Technically, we can understand intermediate asymptotics as follows: Suppose we are interested in a certain property of a system for which there are two constant parameters that we believe play a role in determining the nature of that property. If these governing constants, $X_i^{(1)}$ and $X_i^{(2)}$, have the dimensions of a certain independent variable x_i that also appears in the problem, then an intermediate asymptotics is an asymptotic representation of the property of interest as $x_i/X_i^{(1)} \to 0$ while $x_i/X_i^{(2)} \to \infty$ (Barenblatt, 1996, p. 90).

I will now discuss two relatively simple, but still very technical examples. Both might be grouped under the heading of modeling the spreading of something (heat in the first example, and water in the second) through some medium. I am interested in finding the principal laws and basic features that determine the nature of the spreading in the intermediate asymptotic regime.

Heat

In this section I will show, using a relatively simple example, how it is possible that the investigation of an appropriate asymptotic regime can yield solutions describing the salient features of a phenomenon of interest. I will begin by considering the heat diffusion equation in one dimension.[7] Consider an idealized

---

[6]I am virtually ignorant of theories of art interpretation. So nothing here is meant to deny that part of "understanding" the painting is appreciating the details of *how* it was painted. Nevertheless, there does seem to be a range of distances at which the picture is best discerned.

[7]The discussion that follows is taken largely from discussions in Barenblatt (1996) and Goldenfeld et al. (1989).

Asymptotic Explanation

situation: Suppose that we have an infinite metal bar of circular cross section which has a thermal diffusivity of κ.[8] We are interested in the temperature distribution $\theta(x,t)$ in the bar at different places x and times t after some amount of heat has been locally introduced at some point $x = 0$ (say, by holding a match or lighter underneath the bar at that point). The equation governing this heat conduction is the following partial differential equation:

$$\frac{\partial \theta(x,t)}{\partial t} = \frac{1}{2}\kappa \frac{\partial^2 \theta(x,t)}{\partial x^2} \qquad -\infty < x < \infty. \tag{4.1}$$

We assume that the initial temperature distribution is given by the following Gaussian initial condition:

$$\theta(x,0) = \frac{M_0}{\sqrt{2\pi l^2}} e^{-\frac{x^2}{2l^2}}, \tag{4.2}$$

where l is the width of the initial distribution and M_0 is the "mass" of the initial distribution. That is, M_0 is the area under the initial distribution (4.2). It is the initial amount of heat released into the bar and satisfies the following equation:

$$M_0 = \int_{-\infty}^{\infty} \theta(x,t)dx \qquad t \geq 0. \tag{4.3}$$

The heat equation (4.1) is derived from a conservation law, so that the "mass" M_0 remains constant *for all times t*.

Now the *exact* solution to equation (4.1), given initial condition (4.2), is given by the following equation:

$$\theta(x,t) = \frac{M_0}{\sqrt{2\pi(\kappa t + l^2)}} e^{-\frac{x^2}{2(\kappa t + l^2)}}. \tag{4.4}$$

However, we are interested in the long time $t \gg \frac{l^2}{\kappa}$ behavior of the solution. In this asymptotic regime (4.4) becomes

$$\theta(x,t) \sim \frac{M_0}{\sqrt{2\pi(\kappa t)}} e^{-\frac{x^2}{2(\kappa t)}} \qquad (t \to \infty). \tag{4.5}$$

Note that the asymptotic solution (4.5) lacks a parameter that appeared in the full solution (4.4): namely, the width l of the initial distribution. As a result, we can also arrive at the solution (4.5) by letting $l \to 0$ keeping t fixed. In this degenerate limit, the initial condition becomes a delta function:

$$\theta(x,0) = M_0 \delta(x). \tag{4.6}$$

This means that the solution (4.5) to the long-time asymptotic behavior of the heat equation for a wide variety of different initial conditions—initial distributions of different widths l—is given by the special degenerate solution with initial condition given by (4.6).

---

[8] κ is a property of the material of which the bar is made.

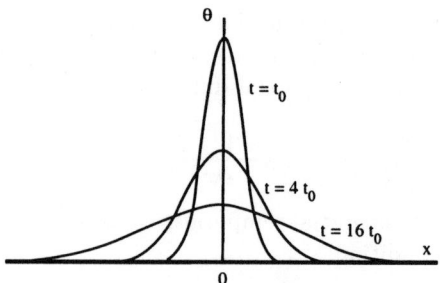

Figure 4.4: θ vs. x

Recall Goldenfeld et al.'s (1989, p. 359) remark that "[i]f one is solely interested in studying universal macroscopic properties of a system, one can choose the most convenient or simplest (in some sense) microscopic model." In the current context—where instead of large-system limiting behavior we are concerned with long-time behavior—this means that we might as well use the delta function initial condition. The long-time behavior of solutions to the heat equation is insensitive to the details of how the heat was introduced into the bar. The "macroscopic phenomenology" (that is, the "shape" of long-time heat distributions) does not depend on details at the "microscopic level" (the "shape" of the initial distribution). We can see that very special idealized initial conditions have much broader relevance than is often thought.

The solution (4.5) possesses the important property of *self-similarity*, which means that the temperature distributions along the bar at different times can be obtained from one another by a similarity or scaling transformation.[9] The θ vs. x plot of (4.5) for different times is given in figure 4.4. Moreover, self-similarity entails that when plotted in the "self-similar" (dimensionless) coordinates $(x/\sqrt{kt}, \theta\sqrt{kt}/M_0)$, each of these curves is represented by a single curve as shown in figure 4.5.

Let us consider the problem of obtaining the temperature distributions at various locations and times from a different perspective—pure dimensional analysis. Suppose that all we know is that the temperature at point x at time t should be a function of the time t, the distance x from the source, the thermal diffusivity κ, the width l of the initial distribution, and the "mass" or quantity of heat M_0:[10]

$$\theta = f(x, t, \kappa, l, M_0). \tag{4.7}$$

If this were known, then even without knowing that the phenomenon was governed by the heat equation (4.1), we can arrive at a solution having the same

---

[9]Let $u(x,t)$ be some observable that is a function of space and time. A similarity transformation has the following form: $u(x,t) \sim t^\alpha f(xt^\beta)$, as $t \to \infty$ for some function f.

[10]This is just the perspective from which we analyzed the period of the pendulum in section 2.2.

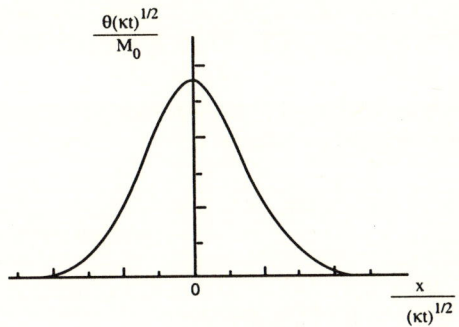

Figure 4.5: Plot in "self-similar" coordinates

form as the self-similar solution (4.5).

This can be accomplished using dimensional analysis. Here is another case in which much information can be gained despite having only limited background theoretical information. Arguments of this form play a major role in the development of mathematical models.[11] Dimensional analysis, as I emphasized earlier, is a form of asymptotic reasoning—a fact I believe is not that widely recognized.

To do dimensional analysis on this problem, first choose a system of units (θ = degrees Celsius, meters, seconds). This specific choice of units is, of course, largely arbitrary. What matters is that it is a member of the class of units (Θ, L, T), where Θ is the dimension for temperature and is independent of the dimensions L—of length, and T—of time). Then the dimensions of the various quantities assumed to be involved in the problem (the variables and parameters) are:[12]

$$[\theta] = \Theta; \quad [x] = [l] = L; \quad [t] = T; \quad [M_0] = \Theta L; \quad [\kappa] = L^2 T^{-1}. \tag{4.8}$$

Note next that the dimensions of the variable t and the parameters κ and M_0 are independent of one another—neither can be expressed as products of powers of the dimensions of the other two. On the other hand, the dimensions of the remaining variables and parameters *can* be expressed as products of powers of these three independent dimensional quantities:

$$\begin{aligned} [x] &= [t]^{1/2}[\kappa]^{1/2}[M_0]^0 \\ [l] &= [t]^{1/2}[\kappa]^{1/2}[M_0]^0 \\ [\theta] &= [t]^{-1/2}[\kappa]^{-1/2}[M_0]^1. \end{aligned}$$

---

[11] It is important to note again that the simplification provided by dimensional analysis does not work for all, or even most, problems. The reasons for this are crucial, having to do with the singular nature of the limiting situation. They will be considered further in section 4.3.

[12] Remember that "[•]" $\stackrel{\text{def}}{=}$ "the dimension of •."

Now form the dimensionless parameters:

$$\Pi = \frac{\theta}{M_0(\kappa t)^{-1/2}}; \quad \Pi_1 = \frac{x}{(\kappa t)^{1/2}}; \quad \Pi_2 = \frac{l}{(\kappa t)^{1/2}}. \tag{4.9}$$

Dimensional analysis lets us express the functional relationship (4.7) using the dimensionless Π's in (4.9) as follows:

$$\Pi = F(\Pi_1, \Pi_2) = F(\frac{x}{(\kappa t)^{1/2}}, \frac{l}{(\kappa t)^{1/2}}) \tag{4.10}$$

In the limit $l \to 0$ we have the idealized delta function distribution that corresponds to letting $\Pi_2 \to 0$. Then we have

$$\Pi = F(\Pi_1, 0)$$

and using (4.9) we get

$$\theta(x,t) = \frac{M_0}{(\kappa t)^{1/2}} F_s(\frac{x}{(\kappa t)^{1/2}}). \tag{4.11}$$

So, *by dimensional analysis alone* we have discovered that the solution to our problem has self-similar form with the function F_s, sometimes called the "scaling function," to be determined. Compare the form of (4.11) with that of (4.5). Clearly, we know that the relationship between the plots of figure 4.4 and the plot of figure 4.5 must obtain. The solution is self-similar.

Had we access to the heat equation (4.1) (as we, in fact, do), we could now substitute this form into that equation and find that we can easily solve for F_s. The benefit of dimensional analysis here is that after the substitution we have only to solve an *ordinary* differential equation to find the unknown scaling function F_s rather than the original partial differential equation. The result is:

$$F_s(\xi) = \frac{1}{\sqrt{2\pi}} e^{-\frac{\xi^2}{2}} \quad \text{where} \quad \xi = \frac{x}{\sqrt{\kappa t}}. \tag{4.12}$$

For the special δ-function initial condition in which $l = 0$, namely that given by equation (4.6), we have self-similarity, and F_s is the scaling function for that solution. In other words,

$$\Pi = F_s(\Pi_1).$$

But, for any initial condition in which $l \neq 0$, the solution must be given by

$$\Pi = F(\Pi_1, \Pi_2),$$

which is *not* self-similar. In the previous discussion of the $(t \to \infty)$ limiting solution (4.5), we saw that the two solutions coincided. It was legitimate to assume $\Pi_2 \to 0$ for sufficiently long times. Given this assumption, we see that

$$F(\Pi_1, 0) = F_s(\Pi_1). \tag{4.13}$$

It is this assumption, therefore, that "allows us to determine the long time behavior of the [heat] equation, starting from initial conditions with $l \neq 0$, from the similarity solution" (Goldenfeld et al., 1989, p. 292).

The assumption that $\Pi_2 \to 0$ reflects our "common sense" belief that the heat diffusion process "forgets" the form of the initial distribution once the mean squared distance $\langle x \rangle^2 \gg l$ (Goldenfeld et al., 1989, p. 292). Thus, asymptotic analysis here introduces a kind of time asymmetry not present in the exact heat equation, which is time reversal invariant.

For situations in which similar assumptions are legitimate, one can proceed as follows using dimensional analysis: First, nondimensionalize the problem and form the relevant dimensionless parameters Π and Π_i. There will then be a solution to the problem of the form

$$\Pi = F(\Pi_1, \Pi_2, \Pi_3, \ldots).$$

If the assumption that some $\Pi_i \to 0$, say Π_2, is legitimate (i.e., that in this limit F is nonsingular),[13] then we know that

$$\Pi \sim F(\Pi_1, 0, \Pi_3, \ldots). \tag{4.14}$$

Barenblatt (1996) calls asymptotics where this assumption is valid, "intermediate asymptotics of the first kind." These situations allow for the relatively easy construction of similarity solutions through traditional dimensional analysis. Similarity solutions are important. They are solutions to a degenerate idealized problem—in our case, this was the solution to the heat equation with the special δ-function initial condition given by (4.6). *But these solutions are not just important as solutions to some idealized special case. As we've seen, these self-similar solutions are also solutions to non-self-similar problems—problems where, for example, $l \neq 0$.* The existence of intermediate asymptotics of the first kind demonstrates that the process under investigation has solutions that stabilize in some asymptotic domain. And so the idealized (far from actual) self-similar solutions have genuine physical significance.

One way to better understand this notion of stability is epistemological—though let me stress that this epistemological characterization is really a consequence of a more fundamental, mathematical conception of stability. Note that within the asymptotic regime, $t \to \infty$, it is not possible to retrodict the actual history of the temperature evolution. That is, one cannot decide whether[14]

> (i) u was a delta function at $t = 0$, or (ii) u was a Gaussian distribution with width l at $t = 0$, or (iii) u was a Gaussian distribution with width l at $t = 10$. These and many other histories all give rise to the same distribution $u(x)$ [namely, that given by equation (4.5) or equation (4.11)] at sufficiently long times, i.e, in the intermediate asymptotic regime. (Goldenfeld et al., 1989, p. 304)

---

[13]This is something that we virtually never know in advance. Where it fails, we may have intermediate asymptotics of the second kind, or no self-similar solutions at all. I will have more to say about intermediate asymptotics of the second kind in section 4.3.

[14]In my example the function u in this passage is the temperature distribution θ.

In finding these self-similar solutions, we find Barenblatt's principal laws and basic features. Intermediate asymptotics of the first kind—dimensional analysis—is just the simplest kind of asymptotic investigation leading to such principal laws and basic features.

It is clear how poorly the causal-mechanical accounts of explanation would fare in explaining the emergence of this long-time asymptotic behavior. On those accounts, one must exhibit the ideal explanatory text—all the details—of the evolution of the temperature distribution in the bar. Given any one initial condition, the amount of computational effort involved in providing the details of the initial transient behavior and of how these transients subsequently decay will be enormous. However, even forgetting about these "practical" considerations, such causal-mechanical explanations will become lost in the details of the distinct transient behaviors. Such phenomena are not repeatable, and their explanations can contribute nothing to an understanding of why the long-time behavior for each system will be the same. That is, each distinct initial distribution will give rise to different transient behavior and, therefore, to completely different explanations.

Water

We have just seen how intermediate asymptotics of the first kind can lead to the relatively simple discovery of the principal features of a phenomenon being modeled. Unfortunately, as I have hinted, the applicability of simple dimensional analysis and the existence of this type of asymptotics is the exception rather than the rule. In most situations, the assumption that justifies equation (4.14) is illegitimate. In other words, more often than not, the aim to simplify via "reduction" by neglecting small parameters leads to a singular problem. A. C. Fowler (1997) argues that a major part of applied mathematical modeling depends on this sort of asymptotic reasoning (he calls it simplification by "reduction"). As the following passage indicates, he expresses a view about detailed numerical "solutions" and their explanatory value consonant with that I have been developing.

> Reduction is the process whereby a model is simplified, most often by the neglect of various small terms. In some cases, their neglect leads to what are called *singular perturbations* ..., whose effect can be understood by using the method of *matched asymptotic expansions*. In particular, it is often possible to break down a complicated model into simpler constituent processes, which, for example, operate on different space and time scales. Analytic dissection in this way leads to an overall understanding that is not so simply available through numerical computations, and indeed it also provides a methodology for simplifying such computations. (Fowler, 1997, p. 7)

In fact, the most interesting cases are cases for which the perturbation problem is singular.

Asymptotic Explanation

For such problems *dimensional analysis* will completely fail to yield a self-similar solution. If, despite this, it is still possible to find some sort of self-similar solution, then we are dealing with what Barenblatt calls "intermediate asymptotics of the second kind."

An apparently slight modification to the heat diffusion problem leads to just such a situation. As we will see, the problem is that the $l \to 0$ limit is not regular. The length l appearing in the initial distribution affects the nature of the solution, no matter how small. It represents another "essential length" in the problem.

There is a very deep analogy, explicitly discussed by both Goldenfeld (1989) and Barenblatt (1996), between the situation here and necessity for a second (typically microscopic) length scale in the renormalization group arguments one finds in quantum field theory and statistical mechanics. My example, of the RG explanation of the universality of critical phenomena discussed in section 4.1, is a case in point.

Let me briefly describe the modification to the problem discussed and then consider some of its consequences. Once again, this is one of the simplest examples that has been rigorously investigated. Consider the problem of describing the shape (height vs. radial distance) of a mound of water filtering through a porous rock medium such as sandstone, over an impermeable (nonporous) bedrock. We assume that the groundwater mound is axially symmetric so that all we care about is how the height of the mound changes over time as the mound spreads out through the ground in a radial direction r.

Figure 4.6 gives some of the relevant details. Assume for simplicity that in the initial mound of water, all of the rock pores are full, and that outside of the initial mound they are all empty of water, though filled with gas. Under the influence of the earth's gravity, the mound of water will begin to spread out and lose height as in figure 4.6. Thus, before the final equilibrium state of uniform (zero) height is reached in the distant future, there will be a time dependent

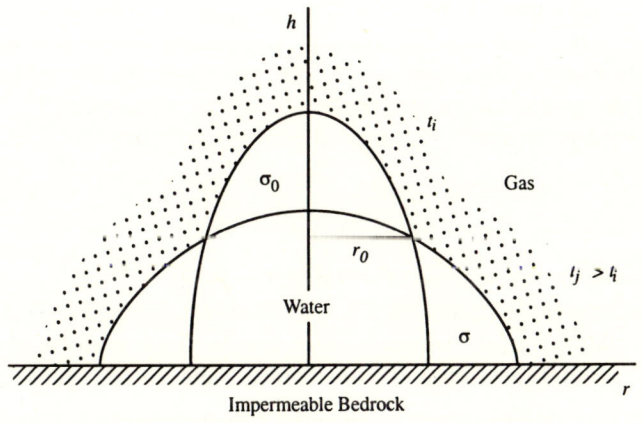

Figure 4.6: Spreading of a groundwater mound

radius, $r_0(t)$, such that for $r < r_0$ the height of the mound will be decreasing with time, and for $r > r_0$ the mound height will be increasing with time:

$$\frac{\partial h}{\partial t} < 0 \quad (r < r_0)$$
$$\frac{\partial h}{\partial t} > 0 \quad (r > r_0). \quad (4.15)$$

For $r = r_0$ it is clear that $\partial h/\partial t = 0$.

Note, also, that as water enters a pore in the rock, only a fraction σ of the pore will become filled—not all of the gas will be displaced. Likewise, as water leaves a pore, there remains, because of capillary forces, a thin layer of water. So a fraction, σ_0, of water is left behind. The equation describing the change in height over time is:[15]

$$\frac{\partial h}{\partial t} = \begin{cases} \frac{\kappa_0}{r}\frac{\partial}{\partial r}\left(r\frac{\partial}{\partial r}\left(h^2\right)\right) & (\frac{\partial h}{\partial t} \leq 0) \\ \frac{\kappa}{r}\frac{\partial}{\partial r}\left(r\frac{\partial}{\partial r}\left(h^2\right)\right) & (\frac{\partial h}{\partial t} \geq 0). \end{cases} \quad (4.16)$$

Equation (4.16) is a special case (in cylindrical coordinates) of a more general equation called the "modified porous medium equation" (Goldenfeld et al., 1989, p. 295).

$$\frac{\partial}{\partial t}u(x,t) = D\kappa \frac{\partial^2}{\partial x^2}u(x,t), \quad (4.17)$$

where

$$D = \begin{cases} 1/2 & \frac{\partial u}{\partial t} > 0 \\ (1+\epsilon)/2 & \frac{\partial u}{\partial t} < 0. \end{cases} \quad (4.18)$$

It is intimately related to the heat equation, in that for $\sigma_0 = 0$—or equivalently, $\epsilon = 0$—the one-dimensional modified porous medium equation (4.17) reduces to the heat equation (4.1). Thus, if water, upon leaving a pore of the rock, left no residual wetting layer, the equations would be the same and we would have the similarity solution discussed in section 4.3. In the general case, however, the groundwater equation (4.16) or (4.17) is nonlinear, whereas the heat equation is linear. In fact, there is a fundamental asymmetry in this problem that is not present in the heat diffusion scenario. The physics of the situation, and not any artifact of the mathematics, is responsible for the asymmetry.[16]

Let us assume an initial condition similar to (4.2) such that the initial "mass" of u,

$$\int_{-\infty}^{\infty} u(x,0)dx = M_0, \quad (4.19)$$

is proportional to the volume of water in the mound. Note a crucial difference between equations (4.3) and (4.19). *The former holds for all times, whereas the latter holds only for the initial time $t = 0$.* This is to be expected, given that as the groundwater mound spreads out, some water remains behind in the

---
[15] κ and κ_0 are constants depending, respectively, on σ and $\sigma - \sigma_0$.
[16] In particular, the responsibility lies with the fact expressed by equation (4.15) and with the existence of distinct constants σ and σ_0.

Asymptotic Explanation

pores that are being exited and is, therefore, no longer available to fill pores at later times—another consequence of the fundamental difference between the two scenarios.

Without going into the details, were one to try to do dimensional analysis on this problem in analogy with that of the heat equation, one would find *three* dimensionless groups:

$$\Pi = \frac{u}{M_0}(\kappa t)^{1/2}, \Pi_1 = \frac{x}{(\kappa t)^{1/2}}, \Pi_2 = \frac{l}{(\kappa t)^{1/2}}, \Pi_3 = \epsilon,$$

related by

$$\Pi = \Phi(\Pi_1, \Pi_2, \Pi_3).$$

As before, we seek a self-similar solution in the $\Pi_2 \to 0$ limit, hoping to be able to represent the long-time behavior solutions to (4.16) or (4.17) for a wide range of initial conditions of the form (4.2) to an idealized degenerate solution when $l = 0$. Unfortunately, this procedure leads to a contradiction, as both Barenblatt and Goldenfeld demonstrate.

However, if one *postulates* that as $\Pi_2 \to 0$ a solution of the form:[17]

$$\Pi = \Pi_2^\gamma \Phi_s(\frac{\Pi_1}{\Pi_2^\delta}, \Pi_3) \tag{4.20}$$

with γ and δ constants, it *is* possible to find a self-similar solution, despite the failure of conventional dimensional analysis to provide one. In this case we have "asymptotics of the second kind." The exponents γ and δ are completely analogous to the critical exponents that describe the universal behavior of a fluid at its critical point. Renormalization group arguments are used to determine these in the context of critical phenomena, and Goldenfeld et al. (1989) have demonstrated that the exponents assumed to exist in the postulated solution for asymptotics of the second kind can be determined using completely analogous renormalization group arguments. In fact, renormalization group arguments provide the justification for the postulated form (4.20) of the solution.

The self-similar solution has the form:[18]

$$u(x,t) = \frac{M_0}{(\kappa t)^{1/2}} \frac{l^\gamma}{(\kappa t)^{\gamma/2}} \Phi_s(\frac{x}{(\kappa t)^{1/2}}, \epsilon). \tag{4.21}$$

If the limit $\Pi_2 \to 0$ is achieved by letting $l \to 0$, then $u(x,t)$ will be either zero or infinity, depending on the sign of the exponent γ. Hence, equation (4.21) is meaningful only if

$$\lim_{l \to 0} M_0 l^\gamma = \text{constant}. \tag{4.22}$$

Since the only way this can happen is if M_0 depends on l in such a way that in the limit $l \to 0$ the product $M_0 l^\gamma$ remains finite, it follows that the mass of the distribution isn't conserved when $\epsilon \neq 0$ in equation (4.18). This, in turn,

---

[17] See Barenblatt (1996, p. 351).
[18] Note that in this problem $\delta = 0$.

is just a further indication of the fundamental difference between this problem and the temperature distribution problem of section 4.3.

As we saw, in the discussion of the "stability" of the intermediate asymptotic solution to the temperature distribution problem, it is impossible to retrodict, from the form of the asymptotic solution, the nature of the initial situation that gave rise to that solution. Nevertheless, since the "mass" of the distribution in *that* problem is conserved over time, one knows the mass, call this $m(0) = M_0$, of the initial distribution given the mass of the distribution, $m(t)$, in the asymptotic regime $t \gg 0$. They are equal. In this sense, we might say that the initial mass of heat remains "observable" even in the asymptotic domain. However, in the current groundwater problem, M_0 must be a function of the initial distribution width l for equation (4.22) to be satisfied. The inability to retrodict is, therefore, even more pronounced—the initial mass $m(0)$ is "unobservable" in principle. One cannot determine it from knowledge of the mass, $m(t)$, for any time $t \gg 0$.

Furthermore, in the limit $l \to 0$ the initial mass diverges. This singular limit is a reflection of the fundamental physics of the situation and is not an artifact of the mathematical model. The singular nature of limit in these situations will often give rise to divergences in various approximation schemes that may be employed to determine the characteristic exponents such as γ. However, these divergences, common in asymptotic approximation schemes (including the singular perturbation theory often employed in the renormalization of quantum field theory and statistical physics) are reflections of features of the approximations employed to deal with a *physically* singular problem. The exact forms of the divergences are not themselves physically significant. In other words, it is true that the "blow ups," or divergences that perturbative approximations engender when trying to determine, say, the values for critical exponents such as γ, are the result of the singular nature of the physics. *But the particular nature or form that these divergences adopt is an artifact of the approximation scheme employed and isn't fully determined by the physics.*

The purpose of this discussion here, despite the technicalities, is really quite simple. Even in cases where dimensional analysis (the simplest kind of asymptotic investigation into the nature of a physical phenomenon) fails, asymptotic investigations may still be able to provide the "principal physical laws" governing the phenomenon. The primary tool in the applied mathematician's toolbox is what Fowler calls "reduction." This just is asymptotic reasoning in one form or another. What I find to be extremely interesting, from the point of view of scientific explanation and understanding, is the close analogy—indeed, the equivalence—between the methods used by applied mathematicians to find self-similar solutions in the long-time regime, and the renormalization group arguments typically employed by physicists in extreme foundational contexts involving micro/macro descriptions.[19] I've argued here and elsewhere Batterman (1998, 2000) that the reasoning involved in renormalization group arguments is ubiquitous, yet this asymptotic reasoning is often overlooked in discussions of explanation by philosophers of science. It involves, essentially, the investigation

---

[19] See, for example, Goldenfeld et al. (1989).

of the physical problem in an asymptotic regime and results in the exhibition of features emergent in that regime whose mathematical representations are stable under certain sorts of perturbations or fundamental changes.

4.4 Conclusion: The Role of Stability

We have seen that in striving to understand some observed regularity in a physical system, one looks to display the "principal laws" and basic features that underlie that phenomenon. The aim of asymptotic reasoning is to do just this. As Barenblatt notes, one feature of crucial importance is the reproducible observability of the phenomenon of interest. In other words, any experiment in which that phenomenon is manifest must be repeatable.

Of course, this does not mean that there are no observable results that are not reproducibly observable. Many physical systems surely exhibit observable behavior that is not reproducible. Consider the repeated measurement for some observable quantity on a given system. Typically, one will find a spread of values hopefully peaked around some one value. If that spread is relatively small, we can take the most probable value to be the value for that quantity.[20] The different exact values actually observed in any given experimental trial likely differ from this most probable value because of various fluctuations that are beyond the control of the experimenter. A mathematical representation or *model* of the system (or at least of that aspect of the system that is under scrutiny) will be a good one, in part, if it is stable in some sense under perturbation of those details responsible for the fluctuations observed.

This idea of stability under perturbation, or structural stability, is extremely important. In a crucial sense it is what leads us to take the principal features resulting from asymptotic analysis to be explanatory. We have already seen explicitly how this works in the case of the the RG explanation of critical phenomena in section 4.1. It is worth elaborating in a more general setting on this issue.

It is true that a certain kind of stability is guaranteed simply by investigating the system and its equations in the intermediate asymptotic regime. If one finds intermediate asymptotic solutions, self-similarity, or scaling laws, then virtually by definition, the observed phenomenon, and the solutions to the governing equations themselves, will be insensitive to various changes to the initial and boundary conditions. But a stronger sort of robustness is also required.

It is surely a goal of scientific explanation to find or "derive" the governing equation for the phenomenon being investigated. Applied mathematicians would call this "constructing a model." In most practical cases, this aspect of mathematical physics is often a matter of educated *guesswork*. It is here that the physicist's experience and intuition play their most prominent roles. However, it is extremely unlikely that the equation one finds is the "real" equation that governs the behavior of the system. Typically, things are much too complicated

---

[20]We can say that that value for the quantity is reproducibly observable, if we, as seems reasonable, are willing to think of this in a statistical sense.

for us to expect this to be the case. Thus, the equation one finds (if, indeed, one finds one at all)[21] is more likely than not an approximation to the "true" or "real" equation. In fact, I'm not really convinced that it makes sense to speak of the "true" governing equation.[22]

Despite this somewhat irrealist attitude, facts about the structural stability of the right aspects of a given equation or model may very well provide evidence that we are on the right track, so to speak. That is, they can tell us that the governing equation and the "solution" we have found is a good one. The idea of a structurally stable solution is, roughly speaking, the idea that because the coefficients of the differential equations we take to govern the behavior of the phenomenon of interest are never determined precisely, the *global structure* of the solutions to those equations ought to be relatively insensitive to small changes in the forms or values of those coefficients.

As noted though, not every behavior of a system will be reproducibly observable. Hence, one should not expect every solution of the equation to possess this kind of insensitivity to perturbations of the equation's form.[23] Chen et al. believe, rightly on my view, that

> [i]f there are unstable or irreproducible aspects in the actual system being modeled, then a good mathematical model of the system must have features unstable with respect to the perturbation corresponding to that causing instability in the actual system. Thus a good model should be structurally stable with respect to the reproducibly observable aspects, but must be unstable with respect to the hard-to-reproduce aspects of the actual system. (Chen et al., 1994, p. 117)

Chen et al. make a distinction between "physically small perturbations of the system" that correspond to small tweaks of the actual physical system—*those that produce only minor effects on its reproducible observables*—and "the corresponding *p-small perturbations* of the model." They "require the structural stability of the model only against *p*-small perturbations." Further, they require of the model only the stability of reproducibly observable features (1994, p. 118).

---

[21] If one is trying without having any governing equations at hand to investigate a problem using the recipe of dimensional analysis or the more general recipe of intermediate asymptotics that may apply when the physics is singular (like that in the ground water problem), then the robustness being discussed here is also necessary.

[22] Recall the discussion in section 4.3 about the nonuniqueness of a microscopic description underlying a given macroscopic phenomenology.

[23] When formulating the original conception of structural stability, Andronov and Pontryagin (1937) argued that only structurally stable models will make good potential models for predictive and explanatory purposes. This idea, backed by an amazing theorem of global analysis proved by Piexoto (1962), came to have the status of methodological dogma. See Smale (1980) for a good clear discussion of some aspects of structural stability. The "stability dogma" has been rightly questioned of late. See Guckenheimer and Holmes (1983, p. 259). Particularly so, after it was realized that Piexoto's theorem does not generalize to systems—differential equations—on manifolds of dimension greater than two. These considerations lead to the restricted notion of structural stability being discussed and advocated by Chen et al. (1994, pp. 116–118).

This leads to the following conjecture which corresponds to a weakening of the "stability dogma" mentioned in footnote 23.

- *Structural Stability Hypothesis:* For a good model, only structurally stable consequences of the model are reproducibly observable (Chen et al., 1994, p. 118).

Conversely, one must also hold that "solutions structurally stable against p-small perturbations describe reproducibly observable phenomena" (Chen et al., 1994, p. 118) . Of course, they are well aware of the potential tautology when these two claims are asserted. But this is meant to be a definition of a good model and needs to be assessed from that perspective.

There is certainly no guarantee that asymptotic analysis will result in the discovery of structurally stable scaling solutions in an appropriate asymptotic regime. Nevertheless, the existence of such solutions—the principal laws or basic features that explain the universal phenomenon of interest—is a reflection of the fact that the process governing the behavior has stabilized in a very important way in the asymptotic regime where the details of the initial and boundary conditions play no significant role. A demonstration of the restricted structural stability expressed in the stability hypothesis, serves to warrant claims to the effect that our mathematical "guesses" genuinely are describing physical features of the world. We can conclude that one reason (and, perhaps, the essential reason) that asymptotic analyses so often provide physical insight is that they illuminate structurally stable aspects of the phenomenon and its governing equations (known or not).[24] Hence, *we have principled reasons for ignoring many, if not most, of the details that we cannot control.* As a result, we can be confident that our theories adequately mirror or represent the world.

The explanations I have been discussing involve, essentially, methods for extracting structurally stable features of the equations that purportedly model or govern the phenomenon of interest. These features are often emergent in appropriate asymptotic domains, and their perspicuous mathematical representation constitutes an essential component of our understanding of the world.

---

[24] James Woodward (2000) and others have noted the importance of a kind of robustness and stability in their discussions of causality and explanation. Naturally, I feel that these authors are completely right to be focusing on stability as an explanatory virtue. I believe, however, that the present considerations have a different focus and provide an account of explanation that Woodward, for example, seems to ignore. Woodward stresses the importance for explanation of a kind of invariance and robustness that may be present in a given regularity to some degree or other. Thus, he discusses how "nonlaw-like" regularities may, because of their robustness, play crucial explanatory roles. Woodward is not concerned to answer why-questions about the universality or degree of universality of the regularities that he discusses. That is, he does not, as far as I can tell, ask the question why the regularity has the robustness that it has or has it to the degree that it has. I have been concentrating on this question. We have seen that stable structures are likely to emerge in asymptotic limits precisely because the existence of limiting solutions tells us that many of the details of particular cases are irrelevant for the regularity, that is, the phenomenon of interest. Thus our focus on asymptotic reasoning provides at least a partial *justification* for why it is important to focus on the structural stability and robustness in the explanatory contexts upon which Woodward and others have focused.

5

Philosophical Models of Reduction

Section 2.3 introduced the general schema that operates in most philosophical models of reduction. This schema received its most influential statement in the seminal work of Ernest Nagel (1961). Very roughly, the reduction of one theory, T', to another, T, requires the explanation of the laws of T' by the laws of T, where explanation is to be understood in terms of the deductive-nomological model. That is, T' reduces to T if the laws of T' are *derivable* from those of T.

In this chapter I will first run rather quickly over some familiar ground. We will discuss how the Nagelian model was emended and extended to make it better fit what many take to be actual intertheoretic reductions. Then in section 5.2 I will discuss, again rather briefly, the well-known objection to reductionist strategies based on the multiple realizability argument most often associated with Jerry Fodor's (1974) "Special Sciences" article. I will also present and discuss Jaegwon Kim's understanding of the import of multiple realizability. As we will see, Fodor and Kim come to opposite conclusions about what that possibility really shows. Section 5.3 discusses an alternative conception of reduction proposed by Kim. He calls this the "functional model of reduction." On this view, practically all special sciences will turn out to be reducible to lower level physical theory. Once we see the motivations behind Kim's model, we will have a clear understanding of the stark philosophical disagreement between reductionists such as Kim and nonreductive physicalists such as Fodor and Block. Section 5.4 addresses the explanatory question at the heart of this disagreement. Next, in section 5.5 I will present a response to the multiple realizability argument that depends on understanding multiple realizability as a kind of universality. In the concluding section, I will begin to suggest that in the majority of cases of so-called intertheoretic reductions, it is perhaps best to give up on philosophical conceptions of reduction altogether. Perhaps reduction as it has been traditionally understood is just not the right sort of relation to be worrying about. The next chapter will then explore what I take to be the best means for understanding interesting relations between theories.

5.1 Nagelian Reduction

In the *The Structure of Science*, Nagel asserts that "[r]eduction ... is the explanation of a theory or a set of experimental laws established in one area of inquiry, by a theory usually though not invariably formulated for some other domain" (1961, p. 338). Nagel distinguishes two types of reductions on the basis of whether or not the vocabulary of the reduced theory is a subset of the reducing. If it is—that is, if the reduced theory T' contains no descriptive terms not contained in the reducing theory T and the terms of T' are understood to have approximately the same meanings that they have in T, then Nagel calls the reduction of T' by T "homogeneous." In this case, while the reduction may very well be enlightening in various respects, and is part of the "normal development of a science," most people believe that there is nothing terribly special or interesting from a philosophical point of view going on here (Nagel, 1961, p. 339).

Lawrence Sklar (1967, pp. 110–111) points out that, from a historical perspective, this attitude is somewhat naive. The number of actual cases in the history of science where a genuine homogeneous reduction takes place are few and far between. Nagel, himself, used as a paradigm example of homogeneous reduction the reduction of the Galilean laws of falling bodies to Newtonian mechanics. But, as Sklar points out, what actually can be derived from the Newtonian theory are *approximations* to the laws of the reduced Galilean theory. The approximations, of course, are, strictly speaking, *incompatible* with the actual laws and so, despite the fact that no concepts appear in the Galilean theory that do not also appear in the Newtonian theory, there is, strictly speaking, no deductive derivation of the laws of the one from the laws of the other. Hence, again strictly speaking, there is no reduction on the deductive Nagelian model.

One way out of this problem for the proponent of Nagel-type reductions is to make a distinction between explaining a theory (or explaining the laws of a given theory) and explaining it away (Sklar, 1967, pp. 112–113). Thus, we may still speak of reduction if the derivation of the approximations to the reduced theory's laws serves to account for why the reduced theory works as well as it does in its (perhaps more limited) domain of applicability. This is consonant with more sophisticated versions of Nagel-type reductions in which part of the very process of reduction involves revisions to the reduced theory. Let me consider this process as a natural consequence of trying to deal with what Nagel calls "heterogeneous" reductions.

The task of characterizing reduction is more involved when the reduction is heterogeneous—that is, when the reduced theory contains terms or concepts that do not appear in the reducing theory. Nagel takes, as a paradigm example of heterogeneous reduction, the (apparent) reduction of thermodynamics, or at

Philosophical Models of Reduction

least some parts of thermodynamics, to statistical mechanics.[1] For instance, thermodynamics contains the concept of temperature (among others) that is lacking in the reducing theory of statistical mechanics.

Nagel notes that "if the laws of the secondary science [the reduced theory] contain terms that do not occur in the theoretical assumptions of the primary discipline [the reducing theory] ..., the logical derivation of the former from the latter is *prima facie* impossible" (1961, p. 352). As a consequence, Nagel introduces two "necessary formal conditions" required for the reduction to take place:

1. *Connectability* "Assumptions of some kind must be introduced which postulate suitable relations between whatever is signified by 'A' [the offending term in the reduced theory] and traits represented by theoretical terms already present in the primary [reducing] science."

2. *Derivability* "With the help of these additional assumptions, all the laws of the secondary science, including those containing the term 'A,' must be logically derivable from the theoretical premises and their associated coordinating definitions in the primary discipline." (Nagel, 1961, pp. 353–354)

The connectability condition brings with it well-known interpretive problems. Exactly what is, or should be, the status of the "suitable relations," often called bridge "laws" or bridge hypotheses? Are they established by linguistic investigation alone? Are they factual discoveries? If the latter, what sort of necessity do they involve? Are they identity relations that are contingently necessary or will some sort of weaker relation, such as nomic coextensivity, suffice? Much of the philosophical literature on reduction addresses these questions about the status of the bridge laws.[2]

The consideration of certain examples lends plausibility to the idea, prevalent in the literature, that the bridge laws should be considered to express some kind of identity relation. For instance, Sklar notes that the reduction of the "theory" of physical optics to the theory of electromagnetic radiation proceeds by *identifying* one class of entities—light waves—with (part of) another class—electromagnetic radiation. He says "the place of correlatory laws [bridge laws] is taken by empirically established *identifications* of two classes of entities. Light waves are not correlated with electromagnetic waves, for they *are* electromagnetic waves" (1967, p. 120). In fact, if something like Nagelian reduction is going to work, it is generally accepted that the bridge laws should reflect the existence of some kind of synthetic identity.

Kenneth Schaffner calls the bridge laws "reduction functions." He notes too that they must be taken to reflect synthetic identities since, at least initially, they require empirical support for their justification. "Genes were not discovered to

---

[1]That this is a paradigm example of reduction practically has the status of dogma in the philosophical literature on reduction. Unfortunately, this is entirely misguided. See Sklar's *Physics and Chance* (1993) for an extended discussion.

[2]For a good introduction to the subtleties involved, see Sklar (1967, pp. 118–121).

be DNA via the analysis of *meaning*; important and difficult empirical research was required to make such an identification" (Schaffner, 1976, pp. 614–615).

Now one problem facing this sort of account was forcefully presented by Feyerabend in "Explanation, Reduction, and Empiricism" (1962). Consider the term "temperature" as it functions in classical thermodynamics. This term is defined in terms of Carnot cycles and is related to the strict, nonstatistical second law as it appears in that theory. The so-called reduction of classical thermodynamics to statistical mechanics, however, fails to identify or associate *nonstatistical* features in the reducing theory, statistical mechanics, with the nonstatistical concept of temperature as it appears in the reduced theory. How can we genuinely have a reduction, if terms with their meanings fixed by the role they play in the reduced theory get identified with terms having entirely different meanings? Classical thermodynamics is not a statistical theory. The very possibility of finding a reduction function or bridge law that captures the concept of temperature and the role it plays in the thermodynamics seems impossible.

This argument, of course, carries with it a lot of baggage about how meaning accrues to theoretical terms in a theory. However, just by looking at the historical development of thermodynamics, one thing seems fairly clear. It seems that most physicists, now, would accept the idea that our concept of temperature and our conception of other "exact" terms that appear in classical thermodynamics, such as "entropy," need to be modified in light of the connection (reduction?) to statistical mechanics. Textbooks, in fact, typically speak of the theory of "statistical thermodynamics." The very process of "reduction" often leads to a corrected version of the reduced theory.

None of this is new. Schaffner and others have developed sophisticated Nagelian-type schemas for reduction that explicitly try to capture these features of actual theory change. The idea is explicitly to include in the model the "corrected reduced theory" such as statistical thermodynamics. Thus, Schaffner (1976, p. 618) holds that T reduces T' if and only if there is a corrected version of T', call it T'^*, such that

1. The primitive terms of T'^* are associated via reduction functions (or bridge laws) with various terms of T.

2. T'^* is derivable from T when it is supplemented with the reduction functions specified in 1.

3. T'^* corrects T' in that it makes more accurate predictions than does T'.

4. T' is explained by T in that T' and T'^* are *strongly analogous* to one another, and T indicates why T' works as well as it does in its domain of validity.

Much work clearly is being done here by the intuitive conception of "strong analogy" between the reduced theory T' and the corrected reduced theory T'^*. In some cases, as suggested by Nickles and Wimsatt, the conception of strong analogy may find further refinement by appeal to what in section 2.3 I referred to as the "physicists'" sense of reduction. That is, we might explicate the notion of T'''s being strongly analogous to T'^* in terms of the emergence of T' as a parameter now appearing in T'^* approaches a limiting value. Recall equation 2.6:

$$\lim_{\epsilon \to 0} T_f = T_c.$$

Let me postpone a discussion of this "physicists'" sense of reduction until chapter 6. Now, I would like to consider an important objection to the Nagelian model of reduction—an objection that apparently holds even against the more sophisticated versions such as Schaffner's.

5.2 Multiple Realizability

The philosophical literature on reduction, particularly on the reduction of the so-called special sciences (of which psychology is taken to be the paradigm case), is largely concerned with issues about multiple realization. In the context of psychology, this involves the claim that the "upper level" psychological properties or states are potentially realized by (or implemented in) a wide variety of *heterogeneous* lower level physical states. Call this the "Multiple Realization Thesis." The Multiple Realization Thesis has been used by a number of people, most notably by Jerry Fodor (1974), as an essential component in an argument to the effect that the special science cannot be reduced to some lower level science and, ultimately, to physics.

In the context of this discussion this means that no Nagelian or neo-Nagelian derivation of the laws of the special science from the laws of physics in conjunction with bridge laws or reduction functions can be possible.[3] As we will now see, this is because the "fact" of multiple realizability is supposed to disallow the very possibility of the requisite bridge laws. The key to the argument is the *heterogeneity* of the lower level realizers of the upper level special science properties. A simple example, discussed by Kim (1992), will help to show how this is supposed to work.[4]

Kim's example concerns a purported case of multiple realizability in which the mineral jade has two distinct physical realizers: jadeite and nephrite.[5] Jadeite and nephrite are distinct chemical kinds. The issue is whether we can reduce a "theory" about jade to a lower level theory about its chemical realizers. So, suppose it is a law of our jade theory that *all jade is green*.[6] Under the

---

[3] For the purposes of the discussion here, I'm just going to take physics to be the reducing theory. Of course, it may be better to talk about the reduction of psychology to neurophysiology or some other "intermediate" level theory.

[4] The conception of law in use here (and in most discussions in this literature)—namely, laws have the form $(\forall x)(Fx \to Gx)$ with some sort of implied necessity—is, on my view, physically ridiculous. This philosophical abstraction ignores many of the important features of laws, particularly in physics. Of course, my saying this here doesn't mean that I have anything like a decent account of what laws are really like.

[5] This example was first introduced by H. Putnam (1980) in "The Meaning of Meaning."

[6] Kim uses the example to argue against the view that one can genuinely have a theory about jade because of multiple realizability. More about that argument later. For my purposes

assumption that jade has only the two realizers, one might try to construct a bridge law as follows:

- (L) $(\forall x)(x$ is jade $\leftrightarrow (x$ is jadeite $\vee\ x$ is nephrite$))$.

Since, by hypothesis, the extension of jade is exhausted by all instances of jadeite and nephrite, (L) expresses a necessary coextensivity.

Fodor's argument against reductionism proceeds by noting that because the microstructures of jadeite and nephrite are completely distinct and heterogeneous from the point of view of the lower level theory, this bridge law cannot express the coextensivity of a natural kind term of the jade theory with a *natural kind* term of the lower level chemical theory. Fodor's view seems to be that bridge laws must be *laws* and, therefore, must relate natural kinds to natural kinds. But why? Why not take the disjunction (x is jadeite $\vee\ x$ is nephrite) to be a natural kind of the lower level theory? His answer, roughly, is that heterogeneous disjunctions of kinds are not kinds. Two responses should immediately come to mind. First, why can't heterogeneous disjunctions be kinds? Second, why think that the bridge laws need to be *laws* where this means something like universal generalizations that relate kinds to kinds? Kim offers interesting arguments for the conclusion that heterogeneous disjunctions cannot be kinds. Let me consider these first and then address the second response one might have.

As Fodor readily admits, the notions of a natural kind and of law are intimately intertwined. "There is no firm footing here. If we disagree about what is a natural kind, we will probably also disagree about what is a law, and for the same reasons. I don't know how to break out of this circle, but I think that there are interesting things to say about which circle we are in" (Fodor, 1974, p. 102). Under the assumption that jade *is* a natural kind and given that the "law," *jade is green*, has received much empirical confirmation, one will be puzzled about what to say about the following imagined circumstance. Suppose we find out that all confirming instances of this upper level law have been instances of jadeite and not nephrite. Then what grounds would we have for thinking that were we to examine a bit of nephrite, it would indeed be green? Laws are supposed to be projectible. But it seems that, despite its confirmation, the law about jade fails the projectibility test. What grounds this intuition is the idea that jadeite and nephrite are heterogeneous as chemical kinds. Were they not, then we might very well allow for its projectibility to cases of yet to be examined nephrite.

Another argument of Kim's appeals to the individuation of kinds in terms of causal powers. It relies on the following two principles:

- *Principle of Causal Individuation of Kinds*: Kinds in science are individuated on the basis of causal powers; that is, objects and events fall under a kind, or share a property insofar as they have similar causal powers (Kim, 1992, p. 17).

here, let's just assume we have a genuine theory and that this law is a genuine law, and see how multiple realizability might be used to block reduction.

- *Causal Inheritance Principle*: If an upper level property is realized in a system on an occasion in virtue of a "physical realization base P," then the causal powers of that upper level property are identical with the causal powers of P (Kim, 1992, p. 18).

If kinds are individuated on the basis of causal powers, and if the causal powers of the distinct possible realizers for an upper level property like jade are radically distinct (heterogeneous), then the realizations of jade on distinct occasions will be realizations of distinct kinds. Appealing to disjunctive properties as kinds seems besides the point.

Now it must be admitted here that Kim intends his arguments to go much further. In fact, he uses these two arguments to support the conclusion that either (1) the upper level law is not a genuine scientific law, or (2) if we insist that it still is a law, then the upper level theory must be reducible to (in a relatively strong sense meaning, perhaps, "eliminable by") the lower level science.

Briefly, the argument proceeds as follows. We grant that the bridge principle (L) expresses the necessary coextension of jade with the disjunction of jadeite and nephrite. Furthermore, because of their distinct and wildly heterogeneous nature, their disjunction cannot be a natural kind of the lower level theory. (Fodor, presumably, would agree with this since he too holds that the disjunction of the realizers can't be a natural kind.) But then jade is nomically coextensive with a disjunctive nonkind. *Then why isn't jade itself equally disjunctive and hence, nonnomic?* Our assumption that the generalization, *jade is green*, is a genuine scientific law, and that our theory of jade is a genuine scientific theory, is simply mistaken. At best we might say that there are two theories—a theory of jadeite and a theory of nephrite.

Kim puts the argument as follows. Here he is talking about mental properties M as the upper level properties. We can just replace "M" and "mental kinds" everywhere by "jade."

> [I]nstances of M that are realized by the same physical base must be grouped under one kind, since *ex hypothesi* the physical base is a causal kind; and instances of M with different realization bases must be grouped under distinct kinds, since, again *ex hypothesi*, these realization bases are distinct as causal kinds. Given that mental kinds are realized by diverse physical causal kinds, therefore, it follows that mental kinds are not causal kinds, and hence are disqualified as proper scientific kinds. (Kim, 1992, p. 18)

So Kim concludes that the multiple realization thesis forces us to say one of two things. Either the special sciences (with multiply realized properties) are not genuine sciences at all, or if we insist that they are genuine sciences, then they are not autonomous. In fact, they will be reducible to the lower level "theories" of the realizers according to a new model of reduction Kim offers.

Kim's new model of reduction is, in fact, distinct from the Nagelian and neo-Nagelian models. Furthermore, it virtually guarantees that "theories" of the special sciences will reduce to lower level physical theories. He calls this the "functional model of reduction." It purports to be a model that can answer explanatory why-questions about the nature of the properties of the upper level theory without having to come up with bridge laws as do the Nagelian and neo-Nagelian models. It is, however, a kind of reduction that will likely not satisfy the so-called "nonreductive physicalists"—those who want to maintain the autonomy of the special sciences.

5.3 Kim's "Functional Model of Reduction"

The very idea that psychological (or upper level special science properties in general) are multiply realizable emerged from the functionalist account of mental states proposed in the 1970's by Putnam and others. If mental states or properties can be given a functional definition, then it comes as no surprise whatsoever that they might be realized or implemented in different physical states.

Kim points out that one, very fruitful, way of characterizing functional properties is as a species of second-order properties.

> F is a *second-order property* over a set **B** of base (or first-order) properties iff F is the property of having some property P in **B** such that $D(P)$, where D specifies a condition on members of **B**. (Kim, 1998, p. 20)

Functional properties over a base **B** are now defined in terms of the nature of the condition D. In particular, D must specify "causal/nomic" relations among first-order properties. Kim offers the following example of a functional property. A substance has the property, dormativity, "just in case it has a chemical property that causes people to sleep. Both Valium and Seconal have dormativity but in virtue of different first-order (chemical) realizers" (Kim, 1998, pp. 20–21). Thus, dormativity is a functional property on Kim's view that is multiply realized.

Furthermore, on the functionalist account of mental properties as understood by Kim, "[t]o be in a mental state is to be in a state with such-and-such as its typical causes and such-and-such as its typical effects," where the causes and effects are understood as holding between "first-order physical properties (including biological and behavioral properties)" (Kim, 1998, pp. 21).

The odd thing about Kim's model of reduction is its historical relationship to the motivations of the functionalist position. Functionalism, after all, was an implementation of the idea that the physical details don't matter with respect to the definition of psychological states, and in conjunction with Fodor-type multiple realizability arguments, it formed the backbone of the *nonreductive* physicalist position. But for Kim, if a property can be functionalized according to the recipe as a second-order property characterized by certain causal/nomic relations among first-order realizers, then that property *reduces* to the lower level theory of those realizers. Here is how it is supposed to work.

Let M be some upper level, say, mental property. Further, suppose that M, as a matter of fact, has two distinct, heterogeneous physical realizers, P_1 and P_2. (For instance, suppose M is pain and P_1 and P_2 are pain's realizers in humans and reptiles.) We assume, with Kim, that this means that P_1 and P_2 are causally heterogeneous and that, therefore, they are distinct as natural kinds. One first construes M (or reconstrues it, if M had not been taken to be a functional property before) as a second-order property defined by its causal/nomic specification D as discussed before. Thus, M is the property of having some property in the base domain with certain causal features.

> The reduction consists in identifying M with its realizer P_i relative to the species or structure under consideration (also relative to the reference world [the world relative to which nomological possibility is to be judged]). Thus M is P_1 in species 1, P_2 in species 2, and so on. Given that each instance of M has exactly the causal powers of its realizer on that occasion ("the causal inheritance principle"), all the causal/explanatory work done by an instance of M that occurs in virtue of the instantiation of realizer P_1 is done by P_1, and similarly for other instances of M and their realizers .... This contrasts with Nagelian reduction via bridge laws in which M only nomologically correlates with P_1 in species 1 but remains distinct from it, and similarly for P_2, and so on. (Kim, 1998, p. 110)

It is worth elaborating a bit further on Kim's model here. There are really three steps involved in a functional reduction. The first and, according to Kim, the most important step is the functionalization of the property to be reduced. Here's how he presents it in an article entitled "Making Sense of Emergence." Let E be the property that is a candidate for reduction to a reduction base **B**. **B** is the class of properties comprising the "basal" conditions for E.[7] The functional reduction of E to **B** involves the following steps (Kim, 1999, pp. 10–11):

Step 1. "E must be *functionalized*—that is, E must be construed, or reconstrued, as a property defined by its causal/nomic relations to other properties, specifically properties in the reduction base **B**."

Step 2. "Find realizers of E in **B**. If the reduction, or reductive explanation, of a particular instance of E in a given system is wanted, find the particular realizing property P in virtue of which E is instantiated on this occasion in this system."

Step 3. "Find a theory (at the level of **B**) that explains how a given realizer of E performs the causal task that is constitutive of E (i.e., the causal role specified in Step 1). Such a theory may also explain other significant causal/nomic relations in which E plays a role."

[7] In supervenience talk, **B** is the subvenient base for E.

The functionalization of E described in Step 1 involves *defining* E as a second-order property as discussed. Kim notes the scientifically significant (and hard) part of this reduction is comprised in Steps 2 and 3.[8] But, as far as the concept of reduction is concerned, the most important step is the first. Kim's functional model also, as we've seen, is supposed to supply explanations.

Suppose that E can be functionalized. Then one can *explain* why a given system S exhibits E at some time t. The explanation takes the following form: E is the property of having some property in **B** that meets the causal specification D—that is, E is functionalized. S has property E at t because S has property P and P fulfills D—P is a realizer of E. Furthermore, we have a *theory* that tells us exactly how P in **B** fills the causal role D.

Therefore, the functional model of reduction works because it allows for the *identification* of E in a given structure or species with its realizer in that structure or species. Thus, an explanatory question, such as "why does this system have pain now?," is answered by pointing out that the system of kind i is now instantiating P_i which is a realizer of pain in systems of that kind. That is to say, P_i meets the causal/nomic specification which is definitive of having pain.

But notice what has happened here. The reduction afforded by Kim's functionalist model is a species—or structure—specific reduction. We have, for instance, reductions of human psychology to human physiology (and ultimately to physics) and reductions of reptilian "psychology" to reptilian physiology (and ultimately to physics). We do not, however, have a reduction of *psychology* to physics. This, according to Kim, is the price of taking the causal inheritance principle and multiple realizability seriously. Therefore, someone who takes psychology to be an autonomous science on the basis of Fodor-type multiple realizability arguments will be forced, if Kim is right, to give up that very position. In effect, any special science for which the functionalization of its special properties is possible will be eliminated. This is why I noted earlier that Kim's model is not likely to be endorsed by those who want to maintain the orthodoxy expressed by the label "nonreductive physicalism."

It is helpful to think about this in terms of the distinction between types of why-questions discussed in section 3.1. Recall that a type (i) why-question asks for an explanation of why a given instance of a pattern obtained. Type (ii) why-questions ask why, in general, patterns of a given type can be expected to obtain. The type of explanation that Kim's model of reduction supplies are answers only to questions of type (i). It is because of this that he is forced to say that the reductions are species-, or structure-specific, and hence that there can be no general reduction of psychology, say, to lower level theory. The explanation of why a given system is now feeling pain is an account of why a

---

[8]Kim uses the example of the reduction of the gene to the DNA molecule. It took many years of research to determine that DNA is a realizer of the gene (1999, p. 11). On the other hand, it is relatively trivial to functionalize the property of being a gene: "[T]he property of being a gene is the property of having some property (or being a mechanism) that performs a certain causal function, namely that of transmitting phenotypic characteristics from parents to offsprings." (1999, p. 10)

particular instance of pain is occurring in a system at this time. By providing the theory and the account that shows how it is that the particular physical realizer P_i meets the causal/nomic specification that is definitive of having pain in systems of kind i, one is providing something like causal-mechanical details for the occurrence of that particular instance of pain.

On the other hand, once we realize that one might be asking a question concerning why pain, in general, appears in certain circumstances, we see that this account provides a completely inappropriate answer to that question. Thus, if we can think of pain, for example, as being a universal property, then it is legitimate to ask for an explanation of its universality. (See the discussion in section 5.5.) But no answer to this question can be provided by Kim's model of reductive explanation. If such an answer can be provided then one need not, it seems to me, adopt the radical eliminitivist line towards the special sciences that Kim seems upon occasion to endorse.

The next section considers several attempts by nonreductive physicalists to diagnose and respond to Kim's argument. Following that discussion, I will offer a solution that in some respects should be welcomed by all of the disputants in this controversy.

5.4 A Metaphysical Mystery

Several authors, notably Jerry Fodor and Ned Block, have offered diagnoses of the intuition motivating Kim's position here. Fodor, for instance, says that "what's really bugging Kim is ... a metaphysical mystery about functionalism" (1997, p. 159). Here's what he has to say about it:

> The very *existence* of the special sciences testifies to reliable macrolevel regularities that are realized by mechanisms whose physical substance is quite typically heterogeneous .... Damn near everything we know about the world suggests that unimaginably complicated to-ings and fro-ings of bits and pieces at the extreme *micro*level manage somehow to converge on stable *macrolevel* properties.
>
> On the other hand, the 'somehow' really is entirely mysterious, and my guess is that that is what is bugging Kim .... [Kim] doesn't see why there should be (how there could be) [macrolevel regularities] unless, at a minimum, macrolevel kinds are *homogeneous* in respect of their microlevel constitution. Which, however, functionalists in psychology, biology, geology, and elsewhere, keep claiming that they typically aren't. (Fodor, 1997, pp. 160–161)

Fodor, as well, confesses to ignorance about the "somehow" here. In fact, he says that he doesn't even "know how to *think about* why" there could be such macrolevel regularities multiply realized by heterogeneous microlevel properties—why, as he puts it, "there is anything except physics" (1997, p. 161).

For Fodor, then, the metaphysical mystery really is how multiple realizability is possible. If it is possible to provide an explanation from lower level theory of the multiple realizability that is present in the special sciences (or at least in some of them), then the mystery may very well be dispelled (at least for some of them). But multiple realizability, as we have seen, seems to tell us that no reduction (except, perhaps, for the species-/structure-specific reductions Kim talks of) can be had. Hence, on the standard way of thinking about reduction, no explanation of this sort can be forthcoming.

David Papineau is also concerned with the fact that the multiple realizability that (often) accompanies functionalism seems to tell us that there can be no commonality among the disparate and heterogeneous realizers that will yield the desired explanation. He finds this to be "incredible."

> The difficulty I am concerned with arises when some mental state S, which mediates between physical input R and physical output T, is realized by a range of different physical states P_i. The puzzle is: why do all the different P_is which result from R all nevertheless yield the common effect T? (Papineau, 1993, p. 35)

Of course, this is another expression of Fodor's metaphysical mystery about functionalism.

Block, as well, thinks that it would "be amazing if laws of nature imposed *no constraints at all* on what can think or be conscious [or, presumably, on what can feel pain]" (1997, p. 120). He argues that while the realizers of the upper level multiply realized properties are heterogeneous, they are not *completely* heterogeneous. "One way of leading to my disagreement with Kim is to note that these special science kinds are typically not nomically coextensive with completely *heterogeneous* disjunctions of physico-chemical properties" (1997, p. 120). In support of this he appeals to what he calls the "Disney Principle." This is the claim that "laws of nature impose constraints on ways of making something that satisfies a certain description."[9] The idea is that such constraints require some sort of similarity between the distinct realizers of the special science properties, so, the realization base is not as heterogeneous as Kim and others might think.

On my view there is indeed something to the Disney principle. But as an assertion of fact about the world, it stands in need of explanation and justification. Without this explanation, it amounts to little more than a denial of Kim's assertion that the physical realizers of the upper level property are so disparate in physical constitution that the disjunctive property nomically correlated with the upper level property fails to be a genuine kind. In other words, one would like to have some independent account of why the Disney principle holds. Block's intuitions, I believe, are correct. But there ought to be a physically acceptable explanation of why. Simply claiming that it would be "amazing" if things were otherwise is not sufficient.

---

[9]The principle is introduced as follows: "In Walt Disney movies, teacups think and talk, but in the real world, anything that can do those things needs more structure than a teacup" (Block, 1997, p. 120). Whence the name.

In the next section I will argue that it is reasonable to think of cases of multiple realizability as cases of similar behaviors in diverse systems. That is, it is profitable to think of multiple realizability as being a kind of *universality*. Given this, it will be possible to (at least) propose a way of explaining multiple realizability in a physically acceptable way despite the heterogeneity of the realizers. We have already seen how this works in section 4.1 in the context of explaining the universality of critical phenomena, and in the other examples of asymptotic explanation discussed in chapter 4, particularly, in sections 4.3 and 4.3.

5.5 Multiple Realizability as Universality

Let us assume then that certain properties that are part of the subject matter of psychology, like pain, are multiply realized. The question posed by Fodor and the others concerns how this can be possible. Can such multiple realizability be explained in a physically acceptable way? If not, then one's physicalism is, on my view, fairly attenuated—to maintain the autonomy (irreducibility) of the special science (in the face of Kim's reductive strategy), it seems one must just dig in one's heels and assert that it is a brute unexplained fact that "unimaginably complicated to-ings and fro-ings of bits and pieces at the extreme *micro*level manage somehow to converge on stable *macrolevel* properties" (Fodor, 1997, p. 160). This seems to be Fodor's attitude, at least if I understand correctly his claim that he doesn't know why (or even how to think about why) there is anything but physics.

Now if pain is multiply realized by heterogeneous physical realizers, then it is the case that there are systems with completely different "micro" details that exhibit the same behavior—that described by the "laws" of pain psychology. That is, after all, what the multiple realization thesis asserts. Furthermore, given the wildly distinct physical nature of the realizers, any detailed causal-mechanical explanation that shows that a given system or realizer exhibits an instance of the behavioral pattern in question will be disjoint from any other such story for a different system or realizer. This is an indication that the details of the realizer must be largely irrelevant for describing the behavior of interest, namely, the fact that the various systems all conform to the particular "law" of pain psychology.

These features are just the features I identified in section 2.2 as features of universal behavior. There I said that universality is characterized by the following general features:

1. The details of the system (those details that would feature in a complete causal-mechanical explanation of the system's behavior) are largely irrelevant for describing the behavior of interest.

2. Many different systems with completely different "micro" details will exhibit the identical behavior.

The philosophical notion of multiple realizability appears to map nicely onto the physicists' conception of the universality of certain behaviors. I have argued elsewhere Batterman (2000) that this is in fact the case and that, therefore, it might be reasonable to expect that explanatory accounts having the same general strategy as the renormalization group accounts may be applicable in these cases as well. In other words, it is reasonable to expect that there will be some sort of asymptotic explanation of the multiple realizability or universality that is characteristic of the special sciences. Furthermore, as we have seen, this sort of explanation involves demonstrating principled *physical* reasons for why many (most) of the details that legitimately allow us to consider the distinct realizers as "wildly" heterogeneous are irrelevant for their universal or multiply realized behavior.

In chapter 8 I will defend the view that the asymptotic explanations of universality essentially involve appeals to structures that are *emergent* in an asymptotic regime. (Recall the brief discussion in section 2.4.) In many cases, though interestingly and importantly not in all, the asymptotic regime in which this emergence takes place is one in which the limit of many interacting components is involved. In the context of the special sciences, particularly in the context of psychology, it seems reasonable to suppose that the physical structures that are, for example, pain's various realizers are structures (states of the system) composed of many interacting parts. Connectionist models of computational features of the mind, for instance, require large numbers of interacting components. In fact, much of the mathematical apparatus of at least one class of network models is formally similar to the statistical mechanical modeling of lattice systems and spin glasses—just those models often used to represent the microscopic details of fluids and magnets (Churchland and Sejnowski, 1992, chapter 3). The mathematics of such models involves, essentially, the study of systems in the limit of a large number of interacting components. Whether these models exhibit singular behavior in certain limits, and whether such behavior can be associated with the emergence of special mental properties, is something I won't speculate about here. Given the widely accepted supposition that pain, for instance, is genuinely multiply realized, I want to claim only that some sort of asymptotic analysis will allow one to explain that multiple realizability.

Thus, I do not want to suggest that the mathematical apparatus of the renormalization group analysis *itself* is what will provide us with an explanation of the multiple realizability of pain. I would have to know much more about the physics and microstructural makeup of the realizers. But, I do not see how any explanation would go unless it proceeds by demonstrating principled physical reasons for why most of the microstructural details of the realizers are irrelevant for the universal behavior of interest. The underlying physical theory must be able to tell us (through some type of asymptotic analysis) that most (otherwise important) physical details are irrelevant.

Suppose that through something like a renormalization group analysis, physics tells us that the physical parameters α and γ are the only relevant physical parameters for the pain universality class.[10] This would be analogous to the renormalization group analysis of critical behavior telling us that the critical behavior depends only upon the physical dimension (d) of the system and on some symmetry property of the order parameter (n). (Recall the discussion of this explanation in section 4.1.) That is, the various realizers of pain in humans, reptiles, and Martians, say, have these features in common when certain generalizations or regularities about pain are the manifest behaviors of interest observed in these systems. Equivalently, physics has then told us that all the other microdetails that legitimately force us to treat the realizers as heterogeneous are irrelevant. With respect to the universal, multiply realized behavior of interest, we can, in effect, perturb the "microdetails" of humans into those of Martians and reptiles without altering the manifest pain behavior we care about. We would then have an explanation of pain's realizability by wildly diverse realizers in exactly the same way that we explain and understand the universality of critical phenomena.

We would also then have a justification for (and an explanation of) Block's Disney principle. In placing humans, reptiles, and Martians in the same universality class characterized by α and γ, physics will have demonstrated that there are constraints on what systems can realize pain.

If all of this could be accomplished, then it would be possible to say something positive about the apparent stalemate between those, like Kim, who favor a kind of reductionist or eliminitivist approach to the special sciences and those, like Fodor and Block, who would maintain the autonomy of the special sciences. Recall that Kim argues that either the special sciences are not genuine sciences, or to the extent that one takes them to be genuine sciences, they are reducible (in a species- or structure-specific fashion) to lower level physical theory. Fodor and others maintain that the special sciences are genuine sciences and that their multiple realizability entails a failure of reduction and (for Fodor at least) little if any hope of physically explaining the very possibility of their multiple realizability.[11]

If I am right and multiple realizability is best understood as physicists understand universality, then it will be possible to explain the multiple realizability—how the "unimaginably complicated to-ings and fro-ings of bits and pieces at the extreme *micro*level manage somehow to converge on stable *macrolevel* properties"—yet also to maintain the nonreducibility of the special science. This is because asymptotic explanations of the renormalization group ilk do not provide sufficient conditions for the universal behavior.[12]

[10] I use the example of pain here only because it is so common in the philosophical literature. There may, of course, be serious problems with this particular case having to do with qualia. On the other hand, in special sciences for which qualia issues do not arise, the explanation can proceed without such worries.

[11] This talk of "genuine science" here depends, of course, on one's intuitions. By and large, it seems to me that most of the intuitions floating around in this debate aren't sufficiently grounded in examples taken from actual scientific practice. (Note that I didn't say "from genuine science"!) I'm urging that a solution to the puzzle about the multiple realizability of the special sciences is possible, once one steps out of the quagmire of ungrounded intuitions.

[12] See Batterman (2000) for details.

It follows that one can have explanation without reduction, since we can accept the Fodor-type multiple realizability argument against reduction, while still allowing for a physically based explanation of that very multiple realizability. This allows us to maintain the autonomy of the special sciences, contrary to Kim, while at the same time endorsing a more robust—less attenuated—physicalism than Fodor believes we are able to. This accords better with Block's version of nonreductive physicalism in which the Disney principle is a step toward a more robust conception.

5.6 Conclusion

As we have seen, there are severe difficulties involved in providing philosophical models of theory reduction. Nowhere are these problems more evident than in the contentious debates that revolve around the question of the status of the special sciences. In this context we have seen that there is an explanatory strategy within which one might fruitfully address the question of how properties of the special sciences can be heterogeneously multiply realized. In physics, as earlier examples have shown, such explanations are common. In their most abstract form, they involve extracting phenomenological relations among upper level properties by determining features that are stable under perturbation of the microdetails.

Multiple realizability gives us good reasons for thinking that certain important intertheoretic reductions, as typically understood by philosophers, will not be possible. I agree. There are indeed good reasons for thinking of the different realizers as diverse natural kinds. It is often legitimate to treat the different systems as genuinely heterogeneous; but this does, for now familiar reasons, make reduction difficult, if not impossible.

On the other hand, it may be best, in this context as well as others, to give up on the various philosophical models of reduction that require a connection (strong or weak) between kind predicates in the reduced theory and kind predicates in the reducing. A more fruitful approach to understanding the relationships between one theory and another is to investigate asymptotic relations between the theory pairs. In these situations the philosophical models of reduction that require derivational connections demand too much detail and, in some cases, too much precision. Sometimes, what is needed to understand relations between theories are methods for ignoring details. As we've seen, asymptotic reasoning is typically the essential component of such methods. Asymptotic methods often allow for the understanding of emergent structures that dominate observably repeatable behavior in the limiting domain between the theory pairs. The next two chapters will consider this approach to intertheoretic relations in detail.

6

Intertheoretic Relations—Optics

This chapter focuses on asymptotic limiting relationships between two theories of light: The so-called ray theory (or geometrical optics) and the wave theory. The aim here is to demonstrate through a detailed discussion of a concrete example (the rainbow) how fruitful it is to study the asymptotic domain between theories. We will see that there are phenomena—various observable features of the rainbow, in particular—that require for their description aspects of *both* the ray theory and the wave theory. To my mind, from a philosophical point of view, this is an extremely important fact. It means that the reductive programs championed by philosophers actually miss out on a lot of interesting physics. The idea that the wave theory reduces the ray theory in the neo-Nagelian sense of reduction discussed in the last chapter turns out to be completely inappropriate in this context. Furthermore, the physicists' sense of reduction, in which the wave theory is supposed to reduce to the ray theory in some appropriate limit, also fails to apply here. We ought to give up on the notion of reduction in any sense and consider what of philosophical (and physical) interest can be learned by studying the asymptotic limits between theories.

Elsewhere, Batterman (1993, 1995), I have argued that one ought to think that a third theory characterizing the asymptotic borderland between the ray and wave theories exists. This theory might, following Berry, be called "Catastrophe Optics." However, Berry himself, doesn't address the philosophical question of whether this is a genuine theory. He says, "[t]he resulting *catastrophe optics* is unfamiliar and unorthodox, cutting across traditional categories within the subject [of optics]" (Berry and Upstill, 1980, p. 259). There are, obviously, issues concerning the legitimacy of treating these asymptotic borderlands as requiring a new theory. I will raise some of them in section 6.5 and in chapter 8 as well.

While this chapter concentrates on the borderland between two specific theories, I believe much of what I have to say generalizes to other theories related by singular asymptotic limits. In chapter 7 I discuss the singular "correspondence" limit relating quantum and classical mechanics as Planck's constant $\hbar \to 0$.[1] I believe, in fact, that any pair of theories related by such singular limits will give rise to new physics requiring a new explanatory theory of the asymptotic domain. For the moment, though, I will concentrate on the wave and ray theories.

The next section begins by discussing what I have called the physicists' sense of reduction. Following this in section 6.2 I will briefly discuss singular limits. Section 6.3 is the heart of the discussion. Here I consider the relationships between ray theoretic structures and wave phenomena. In section 6.4 I show how certain of these relationships are best understood as instances of universal phenomena. This discussion makes explicit the connections between studying asymptotic intertheoretic relations and my earlier discussions of universality, intermediate asymptotics, and Barenblatt's "principal physical laws."

6.1 "Reduction$_2$"

Earlier in section 2.3 I noted that scientists, particularly physicists, use the term "reduction" in a very different sense than do philosophers. This is the sense in which a typically newer and more refined theory, T_f, is said to reduce to another typically older and "coarser" theory, T_c, as some parameter in T_f takes on a limiting value. The philosophical sense of reduction discussed in the last chapter has it that things are the other way around. That is, the older, coarser theory reduces to the newer more refined theory via some kind of deductive derivation (or approximate derivation) of the laws of the older theory from the laws of the newer theory. Thus, the roles of reduced and reducing theories are inverted.

Thomas Nickles introduced this distinction in an important article entitled "Two Concepts of Intertheoretic Reduction" (1973). He labelled the philosophers' sense "reduction$_1$" and the physicists' sense "reduction$_2$." Though Nickles, himself, doesn't express the relationship of reduction$_2$ in quite this way, I think it is easiest to understand it as satisfying the following schema (equation [2.6]):

$$\lim_{\epsilon \to 0} T_f = T_c. \tag{6.1}$$

One must take the equality here with a grain of salt. In those situations where schema (6.1) can be said to hold, it is likely not the case that every equation or formula from T_f will yield a corresponding equation of T_c.

A paradigm case where a limiting reduction of the form (6.1) rather straightforwardly holds is that of classical Newtonian particle mechanics (NM) and the special theory of relativity (SR). In the limit where $(v/c)^2 \to 0$, SR reduces to NM.[2] Nickles says "epitomizing [the intertheoretic reduction of SR to NM] is

---

[1] See also my discussion in Batterman (1995).
[2] I will drop Nickles's subscript on "reduction" unless the context fails to make obvious which sense of reduction is being considered.

the reduction of the Einsteinian formula for momentum,

$$p = \frac{m_0 v}{\sqrt{1 - \frac{v^2}{c^2}}},$$

where m_0 is the rest mass, to the classical formula $p = m_0 v$ in the limit as $v \to 0$"[3] (Nickles, 1973, p. 182).

This is a regular limit—there are no singularities or "blowups" as the asymptotic limit is approached. As I explained in section 2.3 one way of thinking about this is that the exact solutions for small but nonzero values of $|\epsilon|$ "smoothly [approach] the unperturbed or zeroth-order solution [ϵ set identically equal to zero] as $\epsilon \to 0$." In the case where the limit is *singular*, "the exact solution for $\epsilon = 0$ is *fundamentally different in character* from the 'neighboring' solutions obtained in the limit $\epsilon \to 0$" (Bender and Orszag, 1978, p. 324).

In the current context, one can express the regular nature of the limiting relation in the following way. The fundamental expression appearing in the Lorentz transformations of SR, $\sqrt{1 - (v/c)^2}$, can be expanded in a Taylor series as $1 - 1/2(v/c)^2 - 1/8(v/c)^4 - 1/16(v/c)^6 - \cdots$, so the limit is analytic. This means that (at least some) quantities or expressions of SR can be written as Newtonian or classical quantities plus an expansion of corrections in powers of $(v/c)^2$. So one may think of this relationship between SR and NM as a *regular* perturbation problem.

Examples like this have led some investigators to think of limiting relations as forming a kind of new rule of inference that would allow one to more closely connect the physicists' sense of reduction with that of the philosophers'. Fritz Rohrlich, for example, has argued that NM reduces[1] to SR because the *mathematical framework* of SR reduces[2] to the *mathematical framework* of NM. The idea is that the mathematical framework of NM is "rigorously derived" from that of SR in a "derivation which involves limiting procedures" (Rohrlich, 1988, p. 303). Roughly speaking, for Rohrlich a "coarser" theory is reducible to a "finer" theory in the philosophers' sense of being rigorously deduced from the latter just in case the mathematical framework of the finer theory reduces in the physicists' sense to the mathematical framework of the coarser theory. In such cases, we will have a systematic explication of the idea of "strong analogy" to which Schaffner (section 5.1) appeals in his neo-Nagelian model of philosophical reduction.

The trouble with maintaining that this relationship between the philosophical and "physical" models of reduction holds generally is that far more often than not the limiting relations between the theories are *singular* and not regular. In such situations, schema (6.1) fails to hold. Examples of these singular limiting problems are the subject of this chapter.[4] I will argue that it is best to give up on talk of one theory reducing to another on either sense of "re-

---

[3] It is best to think of this limit as $(v/c)^2 \to 0$ rather than as $v \to 0$ since $(v/c)^2$ is a dimensionless quantity, so this limit will not depend on the units used to measure the velocity.

[4] See also Batterman (1995).

duction." Nevertheless, much is to be learned by studying singular limiting relations between theories—both of physical and philosophical interest.

6.2 Singular Limits

We have already come across a couple of examples of singular limits. In section 4.1 I discussed the example of explaining critical phenomena using the apparatus of the renormalization group. The singular limit involved in this example is the so-called thermodynamic limit of large number N of particles. When the system is at criticality—that is, at its critical point—this limit diverges. This is reflected in the divergence of the correlation length. In this context, we can say that the limit of statistical mechanics as $1/N \to 0$ fails to smoothly approach classical thermodynamics. In fact, we must think of the state of a fluid at its critical point as a singularity of thermodynamics where the reductive schema (6.1) necessarily breaks down.

Likewise, in section 4.3 I discussed the problem of the spreading of a groundwater mound over an impermeable rock bed. There too the physics is singular. Mathematically this is reflected in the fact that intermediate asymptotics of the first kind fails completely, as is evident from the fact that the $\Pi_2 \to 0$ limit blows up. This does not, in any straightforward sense, reflect the failure of a limiting relation between a pair of *theories*; however, the mathematics involved is completely analogous to the problems faced when schema (6.1), which expresses a relation between theories, fails to obtain.

Let me add one more simple example to this list of singular limits before moving on, in the next section, to consider in detail the relationship between two theories of light—the ray theory or the theory of geometrical optics and the wave theory.

Consider a wave of wavelength λ (water, light, or sound) traveling in the x-direction with speed v.[5] This wave is represented by the following function:

$$\psi(x,t) = \cos\left(\frac{2\pi}{\lambda}(x - vt)\right). \tag{6.2}$$

Suppose we are interested in the nature of this wavefunction in the case where the wavelength is very small. That is, let's take the limit as $\lambda \to 0$. This would correspond to our interest in, say, the workings of a telescope that are perfectly well characterized (for many purposes) by a view of light as traveling along rays. Unfortunately (or fortunately), the limit is singular; ψ cannot be expanded in powers of λ—the function ψ is nonanalytic at $\lambda = 0$. In fact, at this limit, ψ takes on every value between -1 and 1 infinitely often, in any finite range of x or t.

In wave theory we are often interested in the superposition of two or more waves. Suppose we superpose two waves traveling along the x-axis in different

---

[5] This is taken from Michael Berry's discussion of singular relations between theories in his "Asymptotics, Singularities and the Reduction of Theories" (1994).

Intertheoretic Relations—Optics

directions at speeds v and $-v$. The resulting wavefunction is

$$\psi = \cos\left(\frac{2\pi}{\lambda}(x - vt)\right) + \cos\left(\frac{2\pi}{\lambda}(x + vt)\right) = 2\cos\left(\frac{2\pi x}{\lambda}\right)\cos\left(\frac{2\pi v}{\lambda}\right). \quad (6.3)$$

If we care about the intensity of this new wave at various locations x, we need to calculate $|\psi|^2$. By averaging over time we will get the intensity of a spatially invariant interference pattern:

$$\langle \psi^2 \rangle_t = 2\cos^2\left(\frac{2\pi x}{\lambda}\right). \quad (6.4)$$

We see here that the intensity of a simple two-wave interference pattern has a

$$\cos^2(1/\lambda) \quad (6.5)$$

singularity in the limit $\lambda \to 0$ ray limit.

There are other shortwave singularities that arise in naturally occurring phenomena. Sometimes, according to the wave theory, waves can reach places where, according to the ray theory, there could be no light. For example, from high school physics we are familiar with the fact that for some angles light will be totally internally reflected at an interface between two media such as glass and air.[6] According to the ray theory, no light can be detected outside this interface. Nevertheless, there are waves outside with an amplitude that decays exponentially with increasing distance from the boundary:

$$\psi \propto e^{\frac{-f(x)}{\lambda}}, \quad (6.6)$$

where f is some function of the distance to the boundary.

An interesting example where both sorts of singularities are present—those of the form (6.5) and those of the form (6.6)—is the rainbow. On the lit side of a rainbow, there is the visible oscillatory structure of the bows, and on the dark side apparently no light is visible. In fact, the rainbow is a very instructive, yet relatively simple example. In the next section I will look more carefully at the relationship(s) between the ray and wave theories. We will see that much about these intertheoretic relations can be learned by examining the case of the rainbow.

6.3 Wave and Ray Theories

The Rainbow and the Airy Integral

Let us begin with an account of certain features of the rainbow that is based on geometrical optics. This is the theory in which light travels along rays that can be reflected or refracted by different media, but in which there are no waves.

---

[6]This phenomenon is exploited in the use of fiber optics.

The questions I want to ask concern how much of the observed features of the rainbow are characterizable using the theory of geometrical optics alone. Where does the arc of the rainbow come from? Can we predict where it will be located? Can we describe the observed structure (intensities and locations) of the various bows that we see? Figure 6.1 represents an ideal (spherical) raindrop. Light from the sun hits the raindrop in parallel rays. Upon encountering the droplet, these rays are partially reflected and partially refracted according to simple geometrical rules. In the figure, light is first refracted as it encounters the droplet, then reflected at the back side of the droplet, and then finally refracted again as it exits back into the air. The various rays come in with specific angles of incidence that are specifiable by the variable s and are scattered back as they exit at various angles C.

Using elementary ray optics, Descartes had already shown that the scattering angle as a function of s is a minimum at $C_0 = 138°$. (In other words, all other rays exiting the droplet will exit at angles $C > 138°$.)[7] Furthermore, if $C > 138°$, there are two rays through every point. If $C < 138°$, there are no rays. This is characteristic of the most elementary of stable caustics—the fold.[8] One can see how the refraction and reflection of the light rays produce the fold caustic.

On the geometrical theory of optics, the caustic is a line on which the intensity of light is, strictly speaking, infinite. That is, it represents a singularity of the ray theory. Figure 6.2 is a blowup of the boxed portion in figure 6.1. In effect, the ray theory can locate (only approximately it turns out) the primary bow of a rainbow—the caustic surface—and can also, by considering multiple internal reflections, locate the secondary bow that is sometimes seen. It is incapable, however, of accounting for the following sort of phenomenon. We sometimes notice the presence of so-called supernumerary bows. These are fainter arcs appearing on the lit side of the main rainbow arc. They are visible in the sky

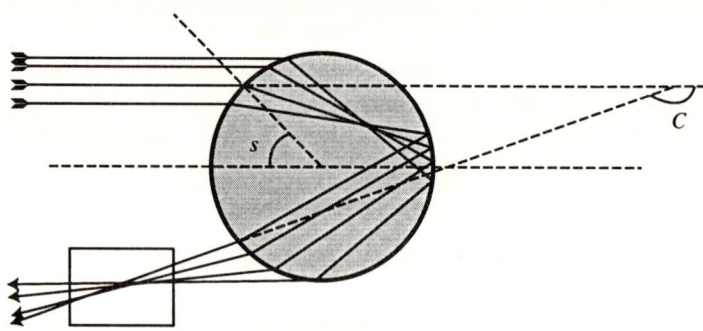

Figure 6.1: Spherical raindrop

---

[7]A very nice discussion of this is provided in Greenler (1980, chapter 1).
[8]Caustics are the envelopes of families of rays. Literally, the term means "burning surface." Figure 6.2 shows the fold caustic.

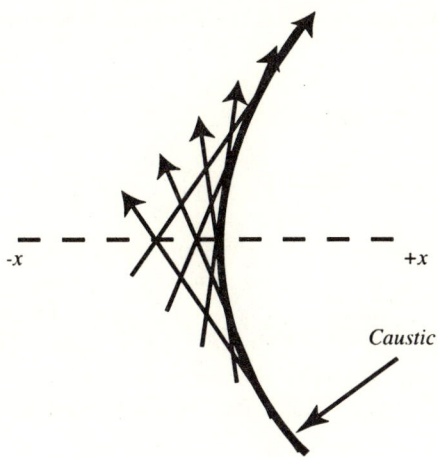

Figure 6.2: The fold caustic

at times as faint alternating pink and green arcs. Furthermore, since actual rainbows do not exhibit light of infinite intensity, the singularity predicted by the ray theory cannot be realistic.

In 1838 George Biddell Airy derived an equation from the undulatory or wave theory of light from which one can determine the variation in intensity of

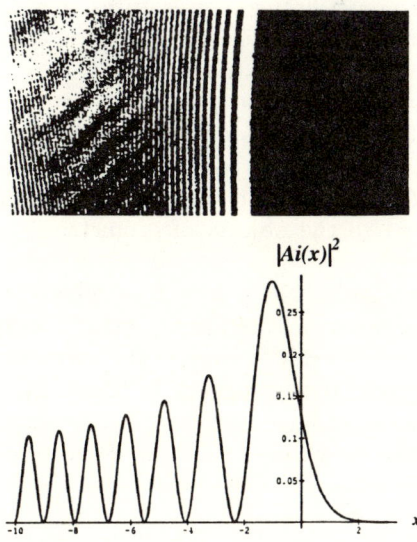

Figure 6.3: Airy function and supernumerary bows

light near a fold caustic of geometrical optics. Airy recognized that the singularity at the caustic was an artifact of the ray theory, and by properly incorporating effects of diffraction using the wave theory, he was able to derive the definite integral of equation (6.7). The square of this, $|Ai(x)|^2$, gives the intensities of the various supernumerary bows of the rainbow caustic and properly represents the smoothing of the ray theory singularity (Airy, 1838, p. 1838).

$$Ai(x) \equiv \frac{1}{\pi} \int_0^\infty \cos\left(\frac{t^3}{3} + xt\right) dt. \qquad (6.7)$$

Figure 6.3 (Berry, 1990, p. 1185) gives an idea of how well the square of $Ai(x)$ can locate and describe the intensities of the wave patterns (the supernumerary bows) decorating the rainbow caustic. It is evident from this figure that the Airy function is oscillatory on the lit side of the caustic (negative values of x) and exponentially decaying in the shadow (positive values of x). It exhibits in the $\lambda \to 0$ limit the singular behaviors characterized by equations (6.5) and (6.6).

Catastrophe Optics

In the context of studying intertheoretic relations, it is reasonable to ask for more details concerning the relationship between the Airy function (6.7) describing various wave features and the "underlying" caustic structure that resides entirely within the ray theory. The function, in effect, describes how (to use a phrase that often appears in this context) the "wave flesh decorates the ray theoretic bones." This colorful way of talking is actually quite suggestive. It is instructive to see exactly how, and to what extent, the ray theoretic structure can be related to, or associated with, certain wave theoretic features. It should come as no surprise that these relations are best studied (and, in fact, can *only* be studied) in the asymptotic domain between the two theories.

Let us see "how much" of the wave phenomena can be captured by ray theoretic considerations. It seems obvious that some sort of connection must exist between the two theories in the asymptotic limit $\lambda \to 0$. What will the waves look like in that asymptotic domain? It seems natural to suppose that they must be related to the objects of geometrical optics—in particular, to families of rays.

At the shortwave limit, where $\lambda = 0$ or where the wavenumber $k = \infty$ ($k = 2\pi/\lambda$), rays and not waves are the carriers of energy. These rays propagate as normals to what we can call "geometrical wavefronts."[9] Consider figure 6.4. We have an initial geometrical wavefront W characterized by its deviation from the plane $z = 0$ by a height function $f(x, y)$. We want to construct, as far as we can, an approximation to a wave at a point P in terms of rays normal to this wavefront.[10]

---

[9]Think of a point source of light such as a candle. The geometrical wavefronts will be concentric spheres with the source as their common center. Light rays will be those straight lines through the center and normal to each sphere's surface.

[10]This discussion follows closely that of Berry (1981) and Berry and Upstill (1980).

Intertheoretic Relations—Optics

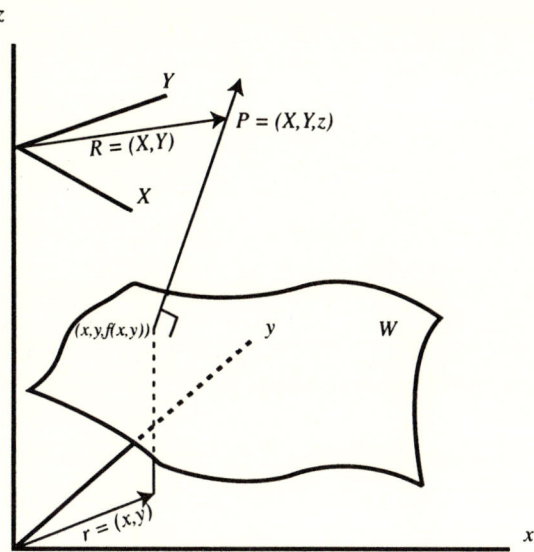

Figure 6.4: Arbitrary wavefront

The rays that emanate from W are defined in terms of the gradient of the so-called "optical distance function" ϕ.[11] To define this function we introduce a second coordinate system X, Y, z (with the same z-axis as the original coordinate system) to label the various points $P = (X, Y, z)$. In this simple case the optical distance function is just the Euclidean distance between the point $(x, y, f(x, y))$ and the point P:

$$\phi(x, y, X, Y, z) = \left[(z - f(x, y))^2 + (X - x)^2 + (Y - y)^2\right]^{1/2}. \qquad (6.8)$$

I assume here that the ray propagates in a homogeneous medium without refraction or reflection from W to P. If we further assume that W deviates only gently from a plane parallel to the plane $z = 0$, then we can use the so-called paraxial approximation to ϕ. (Roughly, this means that since the deviation of the rays from lines parallel the z-axis will be small, we can replace sines of angles by the angles themselves.) On this approximation ϕ reduces to

$$\phi(\mathbf{r}, \mathbf{R}, z) = z - f(\mathbf{r}) + \frac{|\mathbf{R} - \mathbf{r}|}{2z}, \qquad (6.9)$$

where $\mathbf{r} = (x, y)$ and $\mathbf{R} = (X, Y)$.

Following terminology from catastrophe theory, it is useful to think of the points $P = (\mathbf{R}, z)$ as "control" parameters—places, for instance, where we might perform various observations. The points $(\mathbf{r}, f(\mathbf{r}))$ are "state variables"—they label the various starting points for rays that lead to point P.

---

[11] In mechanics this function would be the generating function of a canonical transformation of coordinates. See, for example, Batterman (1995).

Fermat's principle tells us that the rays, for a fixed value of the control parameter **R** are those paths through P for which the optical distance function (6.9) is a minimum:

$$0 = \nabla_{\mathbf{r}}\phi(\mathbf{r}, \mathbf{R}, z) = -\nabla_{\mathbf{r}} f(\mathbf{r}) - \frac{\mathbf{R} - \mathbf{r}}{z}, \tag{6.10}$$

and so

$$\nabla_{\mathbf{r}} f(\mathbf{r}) = \frac{\mathbf{r} - \mathbf{R}}{z}. \tag{6.11}$$

The rays through P are the solutions to this equation. Because of the curvature of the wavefront W, there will generally be more than one such ray though a given point P. (See figure 6.5.) We can label these using an index μ ranging over the state variables or starting points on W:

$$\mathbf{r}^{\mu}(\mathbf{R}, z), \quad \mu = 1, 2, \ldots.$$

Had we started with a wave equation such as the scalar Helmholtz equation

$$\nabla^2 \psi(\mathbf{C}) + k^2 n^2(\mathbf{C}) \psi(\mathbf{C}) = 0, \tag{6.12}$$

it would be natural to assume that as the wavenumber $k \to \infty$ a first approximation to the wavefunction $\psi(\mathbf{C})$ will be a construction out of a superposition of contributions from various rays that pass through the point P.[12] We have just seen that these rays are those that satisfy Fermat's (extremum) condition

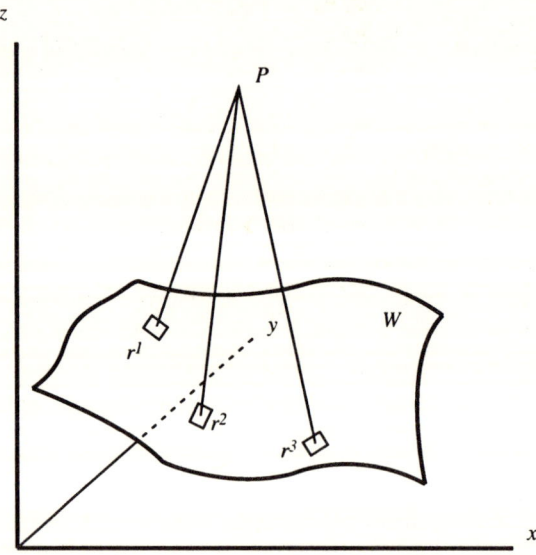

Figure 6.5: Multiple rays through a given point P

---

[12] Here the control parameter $\mathbf{C}$ is the ordered pair $(\mathbf{R}, z)$.

(6.10). If these rays are going to combine somehow to give us an approximation to a wavefunction $\psi(\mathbf{C})$, then we must have a way of determining both the amplitude and phase of ψ in this ray-theoretic approximation.

The phase of the μth ray is 2π times the optical distance ϕ from the wavefront W to $(\mathbf{R}, z)$ in units of wavelength. That is, the phase is

$$k\phi_\mu(\mathbf{R}, z)$$

with $k = 2\pi/\lambda$. We can get an approximation to the wave intensity, $|\text{amplitude}|^2$, by considering the energy flux through an area $dA(\mathbf{R})$ of a tube of rays surrounding the μth ray at the control point $(\mathbf{R}, z)$ when there is unit flux though an area, $dA(\mathbf{r}^\mu)$, surrounding that ray at its starting point on W. (See figure 6.6.) The amplitude is proportional to

$$\left| \frac{dA(\mathbf{r})^\mu}{dA(\mathbf{R})} \right|^{1/2}, \tag{6.13}$$

which is the square root of the Jacobian determinant of the mapping from $d\mathbf{r}^\mu$ to $d\mathbf{R}$:

$$\det \frac{\partial \mathbf{r}^\mu}{\partial \mathbf{R}}(\mathbf{R}, z).$$

Combining these formulas for the amplitude and phase of a given ray though the control point $(\mathbf{R}, z)$ gives the shortwave approximation of equation (6.14).

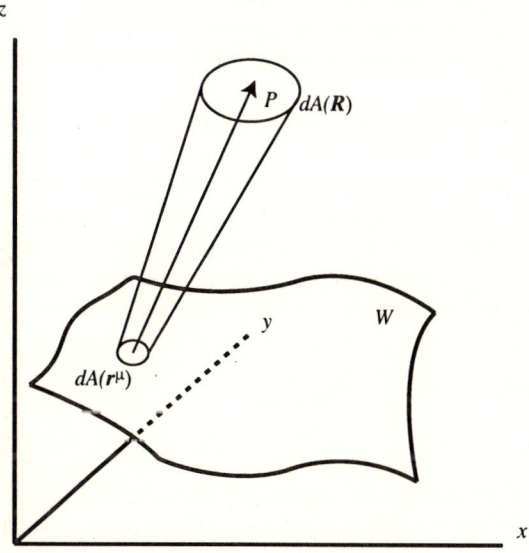

Figure 6.6: "Tube" of rays

Let us call this the "interfering ray sum."

$$\psi(\mathbf{R}, z) \approx \sum_{\mu} \left| \det \left\{ \frac{\partial \mathbf{r}^{\mu}}{\partial \mathbf{R}}(\mathbf{R}, z) \right\} \right|^{1/2} e^{ik\phi_{\mu}(\mathbf{R},z)}. \quad (6.14)$$

Berry notes that

> [a]s a shortwave (or, in quantum mechanics, semiclassical) approximation, [equation (6.14)] has considerable merit. Firstly, it describes the interference between the contributions of the rays through $\mathbf{R}, z$. Secondly, it shows that wave functions ψ are non-analytic in k as $k \to \infty$, so that shortwave approximations cannot be expressed as a series in powers of $1/k$, i.e. deviations from the shortwave limit cannot be obtained by perturbation theory. Thirdly, in the shortwave limit itself $[k = \infty]$, ψ oscillates infinitely fast as $\mathbf{R}$ varies and so can be said to average to zero if there is at least some imprecision in the measurement of $\mathbf{R}$ in the intensity $|\psi|^2$, the terms from [equation (6.14)] with different μ average to zero in this way, leaving the sum of squares of individual ray amplitudes, i.e.
>
> $$|\psi(\mathbf{R}, z)|^2 = \sum_{\mu} \left| \det \left\{ \frac{\partial \mathbf{r}^{\mu}}{\partial \mathbf{R}}(\mathbf{R}, z) \right\} \right|, \quad \text{when } k = \infty \quad (6.15)$$
>
> and this of course does not involve k. (Berry, 1981, pp. 519–520)

The second and third points here tell us, in effect, that the interference effects—the wave features—do not go away in the limit as $k \to \infty$. Thus, as we already knew, there is a qualitative distinction between the limiting behavior and the behavior in the limit.

Despite these advantages, the interfering ray sum (6.14) cannot describe the nature of the wave near and on a caustic. *In other words, it breaks down exactly where we need it most.* It is relatively easy to see what is responsible for this failure. Consider a ray tube surrounding a ray that eventually touches a caustic, say, a fold caustic, such as the one illustrated in figure 6.7. The area $dA(\mathbf{R})$ of the ray tube around the ray $\mathbf{r}^{\mu}$ shrinks to zero on the caustic. Therefore, the amplitude that is proportional (equation [6.13]) to

$$\left| \frac{dA(\mathbf{r}^{\mu})}{dA(\mathbf{R})} \right|^{1/2}$$

becomes infinite on the caustic. Caustics and focal points are the primary objects of study in optics, but the interfering ray sum (6.14) fails exactly at those places. The interfering ray sum can tell us nothing about how the intensity, $|\psi|^2$, on and near the caustic changes as $k \to \infty$, nor can it tells us anything about the wave pattern—the fringe spacings—in that neighborhood.

Now, the construction just considered proceeded from the ray theory "on up" to the wave theory. Most discussions of shortwave asymptotics proceed by

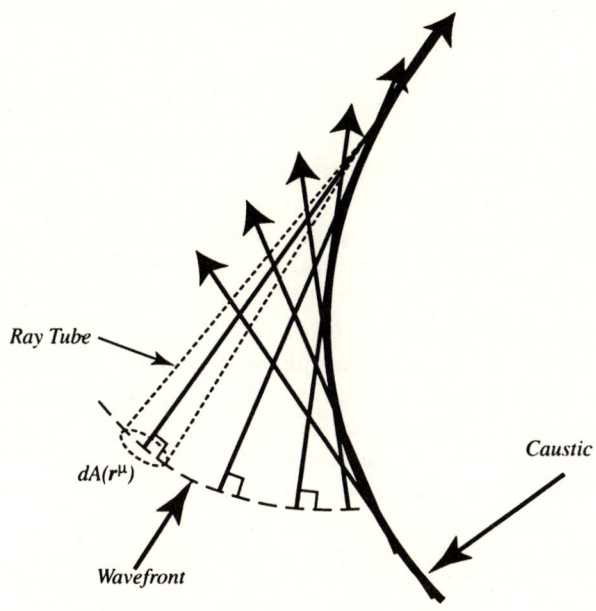

Figure 6.7: Ray tube and a caustic

studying the large k asymptotics of the wave equation (6.12). The first-order approximation of solutions to this equation will, in fact, yield the interfering ray sum.[13] Chapter 7 will provide a fairly detailed discussion of this "top down" asymptotic analysis in the context of quantum mechanics. For now let us just note that there is agreement between the "bottom up" construction I have just presented and a "top down" analysis of the large k asymptotics of the fundamental governing equation.

Nevertheless, both approaches yield the interfering ray sum, which, as we have seen, diverges where we need it most—on caustics. Another asymptotic representation without this problem will be necessary if we are to understand the behavior of light at and near the surface of a caustic. A better representation is provided by considering a continuous superposition of ray contributions instead of the countable sum that appears in the ray sum. That is, we move to an integral representation of the wavefunction ψ. Thus, we can write ψ as follows:[14]

$$\psi(\mathbf{R}) = \left(\frac{k}{2\pi i}\right)^{n/2} \int \cdots \int a(\mathbf{r};\mathbf{R}) e^{ik\phi(\mathbf{r};\mathbf{R})} d^n \mathbf{r}. \qquad (6.16)$$

---

[13] See Berry and Upstill (1980) for a brief discussion.

[14] As before, we let $\mathbf{r}$ be general state variables, $\mathbf{R}$ be the control variables (we now incorporate z in $\mathbf{r}$ and $\mathbf{R}$), n is the number of state variables, and a is the amplitude function to be determined.

We need to try to understand this formula and what it can tell us about the wave patterns near caustics.

As I have already noted in my discussion of the interfering ray sum, for large values of k the exponential in the integrand will oscillate extremely rapidly. Thus, contributions to the integral from neighboring values of $\mathbf{r}$ will almost always cancel out as a result of destructive interference. Only those values of $\mathbf{r}$ for which the phase is stationary—that is, for which the derivatives $\partial \phi / \partial r_i$ equal zero—will give a significant contribution to the integral.[15] But those values of $\mathbf{r}$ for which the derivatives $\partial \phi / \partial r_i = 0$ are exactly those that define *rays*. In other words, the condition of stationary phase is just the condition specified earlier in equation (6.10). Therefore, a natural approximation to equation (6.16) can be determined by the so-called method of stationary phase. But *this just yields the interfering ray sum*—equation (6.14).

Actually, the method of stationary phase yields only the interfering ray sum for control parameters that are relatively far from the caustic. As the point $\mathbf{R}$ moves onto a caustic, the method of stationary phase breaks down because the stationary points coalesce. In effect this provides a diagnosis for the failure of the interfering ray sum on the caustics. But once we have the continuous superposition representation, given by equation (6.16), it is possible to solve this problem for control points near and on the caustic. One method, called "Maslov's method," proceeds by sewing together Fourier transforms at the points on the caustics. It relies on the fact that points on caustics in the standard "position space" representation that I've been using will not lie on caustics in the conjugate "momentum space" representation, and vice versa. Thus, one can avoid the singularities (the coalescence of stationary points on the caustic) by judiciously switching from one representation to another when evaluating the integral.[16]

So it is possible to find asymptotic solutions to the wavefunction (6.16) even at points on a caustic. Suppose we are interested in understanding the nature of the wave pattern decorating a given fold caustic. Then it seems that we would need to find the exact optical distance function (or generating function, see note 11), ϕ, for that particular problem. For instance, if the raindrop isn't quite spherical, then it will have a different geometrical wavefront than that generated by a spherical raindrop. In fact, for each (even slightly) different shaped droplet, the function ϕ will necessarily be different. As a result each solution to the wavefunction will involve a completely different detailed analysis employing Maslov's method. On the other hand, we observe, as a matter of empirical fact, that despite the fact that raindrops have different shapes, the *patterns* in intensity and in fringe spacings that decorate the caustics are similar. *These patterns are universal.* (This should remind us, among other things, of the discussion in section 2.1 of Fisher's intuitions about understanding and

---

[15]This is the key underlying the method of asymptotic evaluation of integrals called the "method of stationary phase." The only points giving positive contribution to the integral in the limit $k \to \infty$ are those for which $\partial \phi / \partial r_i = 0$. These are the stationary points. See Bender and Orszag (1978).

[16]I have discussed this in detail elsewhere. See Batterman (1993, 1995).

Intertheoretic Relations—Optics

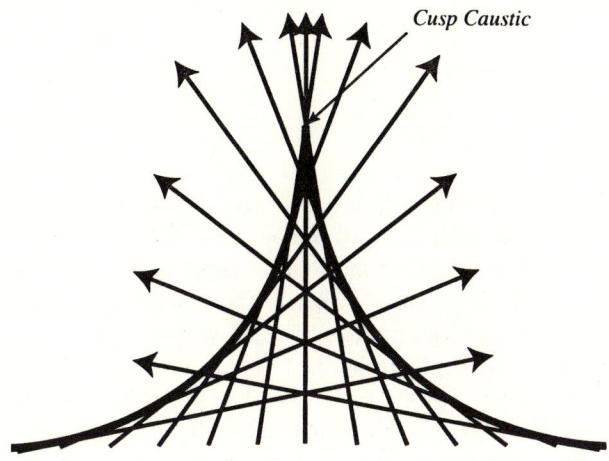

Figure 6.8: Cusp caustic

derivations (explanations) "from-first-principles." Would these detailed individual analyses account for what we are, in fact, interested in—namely, the universal patterns we observe?). There must, therefore, be some structure that is stable under relevant perturbations and such that appeal to that structure will enable us to learn the details of these universal patterns.

Recall from my earlier discussions of universality (sections 2.2 and 4.4) the intimate connection between the existence of universality classes and the structural stability of certain structures. In the present context, the structures are, of course, the caustics. The different caustics are classified, technically, as catastrophes in the sense of Rene Thom's catastrophe theory.[17] Catastrophes, and hence caustics, are distinguished by their codimension K. This is the dimensionality of the control space minus the dimensionality of the singularity. For instance, for the fold caustic in a plane (as in figures 6.2 and 6.7) we see that two dimensions specify control points on the plane (the points at which we might make observations) and the caustic curve is obviously of dimension one. Hence the fold has codimension $K = 1$. (A fold *surface* in three-dimensional control space would also have codimension unity.)

Caustics with $K \leq 4$ are structurally stable in the sense that an arbitrary diffeomorphism (smooth perturbation) will not change its basic form. That is, such perturbations take one member of an equivalence class—a catastrophe— to another member of that same class. For $K \leq 4$, the classes are called the "elementary catastrophes." The elementary catastrophes are representable by distinct polynomials that are linear in the control parameters and nonlinear in

---

[17]See Poston and Stewart (1978) for a good discussion of catastrophe theory and some of its applications.

the state variables. For example, the fold and cusp catastrophes (the first two in the hierarchy) are represented by the following polynomials:[18]

$$\Phi_{\text{fold}}(\mathbf{s}, \mathbf{C}) = \frac{s^3}{3} + Cs \tag{6.17}$$

$$\Phi_{\text{cusp}}(\mathbf{s}, \mathbf{C}) = \frac{s^4}{4} + C_2\frac{s^2}{2} + C_1 s. \tag{6.18}$$

These polynomials are called the "normal forms" of the catastrophes. Figure 6.8 shows a cusp caustic (the point) in two dimensions. These are actually visible in coffee cups that are illuminated at an oblique angle from above by light bulbs.

The importance of the normal forms is the following. Every optical distance function (or generating function) ϕ that yields, say, a fold caustic, can be transformed by diffeomorphism in the control parameter $\mathbf{C}$ into Φ_{fold} as in equation (6.17). Similarly, each ϕ corresponding to a cusp can be transformed into Φ_{cusp} as in equation (6.18). The same holds for the other catastrophes with $K \leq 4$ such as the swallowtail, elliptical umbilic, and the hyperbolic umbilic— Berry and Upstill (1980), Berry (1981).[19] So in order to evaluate the integral (6.16) for any of the infinite number of ψ that encounter a fold caustic, we need only evaluate the "fold diffraction catastrophe" integral. (Similarly for those that encounter cusps, etc.)

The diffraction catastrophe integral for the fold is the following ($C = 0$ on the caustic):[20]

$$\Psi_{\text{fold}}(C) = \frac{1}{\sqrt{2\pi}} \int_{-\infty}^{\infty} e^{i(\frac{s^3}{3} + Cs)} ds. \tag{6.19}$$

Fermat's condition—the ray condition—is that

$$\frac{\partial \Phi_{\text{fold}}}{\partial s} = s^2 + C = 0. \tag{6.20}$$

For $C < 0$ this means that there are two interfering rays $s^\mu = \pm\sqrt{-C}$. For $C > 0$, the roots of (6.20) are complex and there are no real rays. The intensity $|\Psi(C)|^2$ is exactly that given by $\frac{|Ai(C)|^2}{2\pi}$—basically, the square of the Airy function (equation [6.7]).[21]

---

[18] Here, for reasons that will become clear later on, I now let $\mathbf{s}$ and $\mathbf{C}$ denote, respectively, the state variables and the control parameters.

[19] The existence of these normal forms means that all of the generating functions in the same class share the same caustic geometry. In turn, this means that the topology of the coalescence of the stationary points as $\mathbf{C}$ approaches a caustic must be identical.

[20] Note the appearance of the normal form Φ_{fold} (6.17) in the integrand.

[21] A different, but equivalent (for real values) formulation of the Airy equation as a function of complex variables is given by

$$Ai(z) = \frac{1}{2\pi} \int_{-\infty}^{\infty} e^{i\left(\frac{t^3}{3} + zt\right)} dt.$$

Compare this with the fold diffraction catastrophe integral (6.19).

The discussion in this section has led us in a seemingly circuitous route back to the very equation Airy invented to characterize the behavior of light near a fold caustic such as the rainbow. But now we see that this is just the first in a hierarchy of equations—the diffraction catastrophe integrals—that represent different universality classes of wavelike features. But my interest in so-called catastrophe optics was motivated in the first instance by a desire to understand the relationships between two theories, wave theory and ray theory, in the asymptotic shortwave limit. In particular, one question of interest is how the diffraction patterns change in this asymptotic domain as $k \to \infty$. The next section will address this question. There are, in fact, certain exponents that characterize, algebraically, the behavior of both the intensity of light and the way the fringe patterns change with increasing k. These exponents reflect the singularities in the wave physics at the the caustics. In a very deep sense they are analogous to the anomalous exponents that we found earlier in my discussions of critical phenomena in thermodynamics (section 4.1) and in the behavior of the spreading of a groundwater mound over impermeable bedrock (section 4.3).

6.4 Universality: Diffraction Catastrophe Scaling Laws

As $k \to \infty$ the intensity on the caustic must rise toward infinity and the spacing of the fringe patterns must go to zero. The question is how are these limits approached. It turns out that the structural stability of the caustics—*geometrical objects of ray theoretic origin*—guarantees that the k dependence of both the intensity and the fringe spacings is *universal*. In essence what we are looking for is an intermediate asymptotic solution to the physical wavefunction (6.16). Because of the singularities at the caustics we must expect that this is a solution of the "second kind."

In fact, the asymptotic solution is the following scaling (self-similar) law:

$$\psi(R_i) = k^\beta \Psi(k^{\sigma_i} R_i), \qquad (6.21)$$

for the various control parameters R_i ($1 \leq i \leq K$) (Berry and Upstill, 1980, p. 292). The exponents β and σ_i are the "anomalous" or scaling exponents. They will tell us about the nature of the fringe spacings and intensities scale with increasing values of k near the caustics. Let us now try to understand this equation in the simplest case of the fold caustic.

Note first that the wavefunction (6.16), which represents the physical wave we are interested in, depends on k in two places. There is the k dependence in the exponent that multiplies the generating function ϕ (optical distance function) and there is the k appearing in the prefactor, $k^{n/2}$, multiplying the integral itself. The prefactor guarantees that (when we are far from the caustic and hence the integral can be evaluated by the method of stationary phase) the contributions to the interfering ray sum have amplitudes that are k-independent. This is required since the amplitude is determined by the ray tube equation (6.13), which is k-independent as one would expect in the ray theory.

The discussion of the last section 6.3 shows that near the caustic it is possible to transform (perturb) the physical generating function, ϕ, in (6.16) to the normal form of the appropriate catastrophe Φ. For my purposes, I will focus on the fold caustic with its normal form Φ_{fold} (6.17). We also saw that as $k \to \infty$ the wave patterns are described by the diffraction catastrophe integral (6.19). *But this equation is independent of k.* In order to understand the universal way the intensities and fringe spacings scale with increasing k—that is, in order to understand the scaling law (6.21)—we need to understand the relationship between the k-dependent equation (6.16) and the k-independent equation (6.19).

Recall that the normal forms Φ have a part that is nonlinear in the state variables s and independent of the control parameters C. This is called the "germ" of the normal form. The remaining terms of the equations are linear in C and these are called the "unfolding terms." For example, the germ of the Φ_{cusp} (6.18) is $s^4/4$. (See, for example, Poston and Stewart (1978).) We first want to remove the k-dependence of the germ of the function ϕ in (6.16) by a linear k-dependent transformation from r to s: $kr^3 = s^3$. This will yield a factor k^β outside the integral. Second, we remove the k-dependence of the unfolding terms by another linear transformation $R_i = k^{-\sigma_i} R_i$ (Berry, 1981, pp. 538–539) (Berry and Upstill, 1980, pp. 295–296). Since $kr^3 = s^3$ implies that $r = s/k^{1/3}$ and since $dr = k^{-1/3} ds$, we get the equation[22]

$$\Psi_{\text{fold}}(R) = \frac{k^{1/6}}{\sqrt{2\pi}} \int_{-\infty}^{\infty} e^{i\frac{s^3}{3} + k^{\frac{2}{3}} Rs} ds. \qquad (6.22)$$

This is exactly of the form (6.21) with $\beta = 1/6$ and $\sigma_1 = 2/3$. These exponents tell us, respectively, how the intensity and fringe spacings scale with increasing values of k at the caustic. In fact, we see immediately that on the fold caustic the intensity, $|\Psi|^2$ diverges as $(k^{1/6})^2 = k^{1/3}$ and the fringe spacings on the bright side of the caustic shrink as $k^{-2/3}$.

Equations for the fold caustic (6.22) and in general (6.21) are what Barenblatt would call "the principal physical laws" governing the wave features of interest. They are self-similar solutions to the problem of understanding the nature of various wave phenomena in the those situations where the wave intensity is greatest—on the singularities or caustics.

Furthermore, they capture, in a particularly transparent way, certain important intertheoretic relations between the wave theory and the ray theory. In particular, the ray theoretic caustic, with its structural stability properties, is responsible for universal behavior in the intensity and form of various wave patterns. Thus, in the asymptotic domain characterized by the limit $\lambda \to 0$ or $k \to \infty$, *elements from both theories seem to be intimately and inextricably intertwined.* This is a direct result of the singular nature of the limiting relationship.

---

[22]In the case of the fold, caustic the second transformation is unnecessary.

6.5 Conclusion

By focusing on the intertheoretic relations between the ray and wave theories of light using the particular example of the rainbow, I have shown that intertheoretic relations can be much more involved and complicated than the reductive schema 6.1 appears to allow. In fact, it is fair to say that in many (actually in most) cases of asymptotic limiting intertheoretic relations, that schema will fail. Whenever the limiting relations are singular, we can expect new, and often unexpected, physics in the asymptotic regime between the two theories.

Furthermore, it should be evident that the various philosophical accounts of reduction discussed in chapter 5 will fail to account for the physical phenomena that obtain in the asymptotic borderlands. These reductive accounts are, as we've seen, primarily concerned with the derivation of the laws of the reduced theory from those of the reducing. But the kinds of phenomena I have been considering here—the universal behavior of light in the neighborhood of caustics of various types—are neither laws of the reduced nor of the reducing theories. The philosophical views of reduction fail to pay attention to these kinds of issues entirely. Given the discussion of the various ways in which philosophical accounts of explanation cannot deal with universal behavior of certain types, it should not be very surprising that the main accounts of reduction also will fail. After all, these accounts are intimately tied to various views about explanation. We have already taken an important step, in our discussion of multiple realizability in chapter 5, toward seeing how this intimate relationship between reduction and explanation needs to be chaperoned.

So what is the status of *catastrophe optics*—the "theory" of the asymptotic shortwave limit of the wave theory? As Berry says, it cuts across traditional categories in optics. In chapter 7 I will argue that *semiclassical mechanics* plays an analogous role as the "theory" of the asymptotic $\hbar \to 0$ limit of quantum mechanics, cutting across traditional categories of mechanical theories. In what sense are these really *theories* distinct from the pairs to which they are somehow related?

Concerning the diffraction catastrophes that I have been discussing, Berry says the following:

> Such "diffraction catastrophes" have intricate and beautiful structures, and constitute a hierarchy of nonanalyticities, of emergent phenomena par excellence. The patterns inhabit the borderland between the wave and ray theories, because when λ is zero the fringes are too small to see, whereas when λ is too large the overall structure of the pattern cannot be discerned: they are *wave* fringes decorating *ray* singularities. (Berry, 1994, pp. 602–603)

While the visual idiom Berry uses provides an intuitive picture of the character of these properties as emerging in the $\lambda \to 0$ limit, I do not think that it captures the full import of the nature of their emergence. It is true that when $\lambda = 0$ the fringes are too small to discern, but that is because they do not actually exist at that limit. (Recall that this is a reflection of the singular nature of the limit.) The fringes in a sense do exist for any value $\lambda > 0$, but as Berry says, their structure or pattern cannot be observed if the wavelength gets too large. What's crucial, however, is the fact that the nature of the fringe pattern itself—how the spacings and intensities change with changing wavelength—depends upon the structure of the ray theoretic caustic. When the wavelength gets too large, the *wave* patterns can no longer "discern" (or are *no longer dominated by*) the relevant features of the "underlying caustic."

It is true, nevertheless, that these emergent structures are *in some sense contained in* the wave equation (6.12) (and in the case of quantum mechanics, in the Schrödinger equation). The question here is exactly what is the operative sense of "contained in." I think that the sense of "contained in" must refer to asymptotic analysis. It is not the case that these structures are derivable in any straightforward sense from the fundamental wave equation. A theme throughout the book has been how different asymptotic reasoning is from the usual understanding of how solutions to equations are to be gotten. The examples considered in this chapter served to illustrate the fact that asymptotic analysis of singular limiting intertheoretic relations typically yields new structures that are not solutions to the fundamental governing equations in the normal sense of the term.

It is possible to say a bit more about what this sense of being "contained in" but not "predictable from" the fundamental theory amounts to. I have claimed that mathematical asymptotic analysis of the Helmholtz equation (6.12) can yield solutions characterizing the emergent patterns of light on and near the caustics of geometrical optics. (Again, a more detailed quantum mechanical example of this sort of analysis is discussed in chapter 7.) We see that the interfering ray sum (6.14) and the scaling laws result from this asymptotic analysis. *But* we also see that one cannot interpret these purely mathematical results in wave theoretic terms alone. An understanding of what the mathematical asymptotics is telling us requires reference to structures that make sense only in the ray theory. Thus, the interpretation of the interfering ray sum and of the scaling laws demands reference to caustic structures. As we've seen though, these are purely ray theoretic features.

So "predictable from fundamental theory" is somewhat ambiguous. In one sense, the solutions are *contained in the fundamental wave equation*. This is the sense in which asymptotic analysis enables one to find mathematical representations of those solutions. On the other hand, the understanding of those mathematical representations requires reference to structures foreign to the fundamental theory. In this sense, they are *unpredictable from fundamental theory*.

It seems reasonable to consider these asymptotically emergent structures to constitute the ontology of an explanatory "theory," the characterization of which depends essentially on asymptotic analysis and the interpretation of the results. This full characterization must make essential reference to features of both the ray and the wave theories. That is what the analysis forces upon us. So, as we've seen, despite the fact that the wave equation in some sense "contains" these emergent phenomena, their *explanation* necessarily requires reference to features that are not present in the wave theoretic ontology. As a result, I think

Intertheoretic Relations—Optics

it is entirely reasonable to think of catastrophe optics as a genuinely new theory, a third theory, describing the asymptotic "no man's land" between the wave and ray theories.

If these diffraction catastrophe structures and their properties do represent features of a new domain, then perhaps the use of the term "emergent" to describe them is appropriate. Chapter 8 will argue that this is indeed the case.

7

Intertheoretic Relations—Mechanics

Chapter 6 discussed in detail the asymptotic relations between the wave and ray theories of light in the shortwave limit as $\lambda \to 0$. I mentioned at several points that similar relationships obtain between quantum and classical mechanics in the semiclassical limit in which Planck's constant $\hbar \to 0$.[1] This limit too is singular and the result is that a third theory—semiclassical mechanics—characterizes the asymptotic borderland between quantum and classical mechanics. In this chapter I will continue the discussion of asymptotic intertheoretic relations by concentrating on certain aspects of the so-called correspondence between classical and quantum mechanics.

There are truly interesting issues here. Ever since the beginning of the development of the quantum theory, the idea of a correspondence between the two theories in some limit—very roughly, the limit in which "things get big"— has played an important role. Bohr, in fact, required that the "correspondence principle" be a fundamental law of the quantum theory. Jammer puts the point as follows:

> It was due to [the correspondence principle] that the older quantum theory became as complete as classical physics. But a costly price had to be paid. For taking resort to classical physics in order to establish quantum-theoretic predictions, or in other words, constructing a theory whose corroboration depends on premises which conflict with the substance of the theory, is of course a serious inconsistency from the logical point of view. Being fully aware of this difficulty, Bohr attempted repeatedly to show that "the correspondence principle must be regarded purely as a law of the quantum

---

[1] One might reasonably wonder what it could mean to let a *constant* change its value. The way to understand "$\hbar \to 0$" is that it is the limit in which $\hbar$ is small relative to some quantity having the same dimension—namely, the classical action.

theory, which can in no way diminish the contrast between the postulates and electrodynamic theory." (Jammer, 1966, p. 116)

I have argued elsewhere, Batterman (1991, 1993), that semiclassical mechanics is the modern extension of the "old quantum theory" and that many of the issues of concern to Bohr in the early development of the quantum theory still remain. Some of these issues, in a more modern guise, will be addressed here. In particular, I will argue, as in my earlier discussion of catastrophe optics, that semiclassical mechanics should be understood as the explanatory theory of the asymptotic "no man's land" between classical and quantum physics.

7.1 Classical and Quantum Theories

The representation of classical mechanical systems and their evolutions typically utilizes the descriptive apparatus of *phase space*. This is usually a multidimensional Euclidean space in which the complete state of a system is represented by a point. An example is the representation of the state of a simple harmonic oscillator such as a pendulum by a point in a two-dimensional phase space whose axes represent angular position, θ, and angular momentum, ω. (See figure 7.1.) The oscillator has one degree of freedom and the phase space is two dimensional. In general, a system with N degrees of freedom will have its state represented by a point in a $2N$ dimensional phase space. As the system evolves over time according to Hamilton's equations of motion, the representative point carves out a trajectory in this phase space.

The analog of this classical phase space in the optics case discussed in chapter 6 is the space that would be formed by combining (taking the Cartesian product of) the space of state variables s^μ with the space of control variables or parameters **C**. Figure 6.8, for example, shows the cusp catastrophe in the two-dimensional control space (X, Y) with z fixed. The full representation in

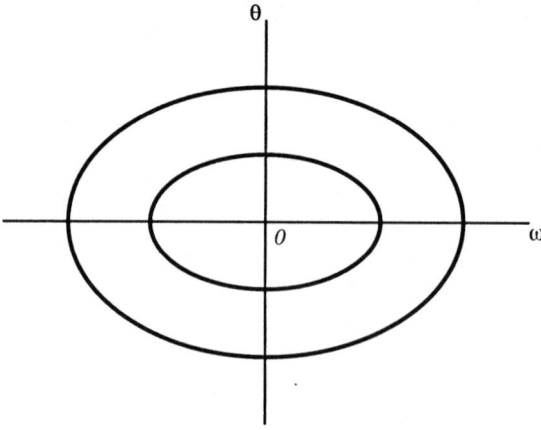

Figure 7.1: Phase space for a simple oscillator

"phase space" is shown in figure 7.2. Recall also equation (6.18). The C_i's are the control parameters.

As we saw in figure 6.8, the caustic organizes the multivaluedness of the family of rays—inside the cusp each point is the intersection of three separate rays, whereas outside there is only one ray through each point. In the combined "phase space" of figure 7.2, this is represented by the foldings of a surface over the control space. In the figure, the three contributions from different sheets of the folded surface are labeled 1, 2, and 3.

As I noted, in classical mechanics the state of a system is represented by a point $(\mathbf{q}, \mathbf{p})$ in phase space. Here $\mathbf{q} = (q_1, \ldots q_N)$ is the generalized coordinate for the system and $\mathbf{p} = (p_1, \ldots p_N)$ is the conjugate generalized momentum. Thus, the classical state is fully represented by $2N$ numbers. In quantum mechanics, on the other hand, the system's state is represented by a wavefunction $\psi(\mathbf{q})$, which is a function of only half the canonical coordinates $(\mathbf{q}, \mathbf{p})$. In a moment I will show how this reduction in the number of coordinates gets carried out. (We are, of course, looking at the relationship between the classical description of a phenomenon and the quantum description.) The "objects" of the classical theory that are relevant for this discussion are families of trajectories or surfaces evolving under Hamilton's equations of motion. These are analogous to the ray families discussed in section 6.3. The analog of the waves in wave optics are, naturally, the quantum mechanical wavefunctions.

We can begin with a question similar to that first asked in my discussion of the asymptotic relationship between the wave and ray theories: How are the classical structures related to or associated with certain quantum mechanical features? How much of the quantum mechanical wavefunction can be under-

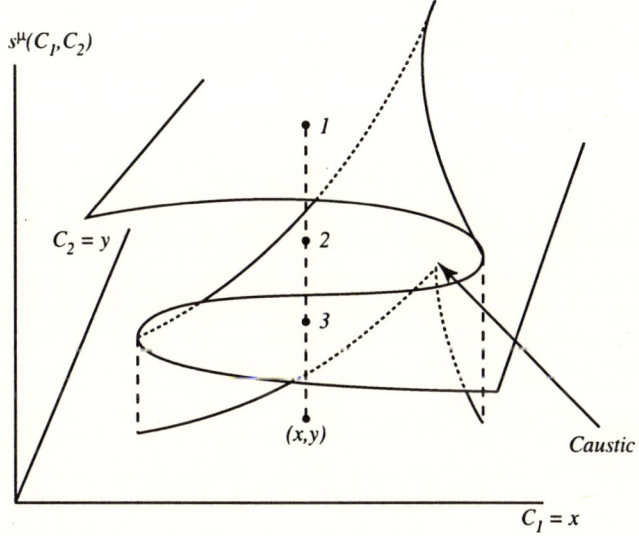

Figure 7.2: Cusp caustic and multivaluedness in "phase space"

stood with purely classical "trajectory-theoretic" considerations?

Given that classical mechanics is supposed to be a limiting case of quantum mechanics as $\hbar \to 0$, we would like to try to understand the nature of the quantum mechanical wavefunctions in this limit. The hope is to find a construction similar to the interfering ray sum (6.14) that we found in section 6.3. In other words, we would like to find an association between wavefunctions and families of phase space trajectories, or equivalently, surfaces in phase space.[2]

The association depends upon geometric properties of certain kinds of N-dimensional surfaces embedded in the $2N$-dimensional phase space. These are so-called Lagrangian surfaces.[3] Consider such a surface Σ embedded in a four-dimensional phase space as in figure 7.3. One can consider points on Σ as a function of the coordinates $\mathbf{q}$—$p(\mathbf{q})$. In so doing we treat the surface as a function of *only half the canonical coordinates* $(\mathbf{q}, \mathbf{p})$, just as the quantum mechanical wavefunction $\psi(\mathbf{q})$. Next we introduce a new set of variables $\mathbf{Q} = (Q_1, Q_2)$ to label points on Σ. (These variables are analogous to the introduction of the coordinates $[X, Y]$ used to locate the point $\mathbf{R}$ in the optical example of section 6.3. Recall figure 6.4.)

Corresponding to this new set of coordinates will be a set of conjugate mo-

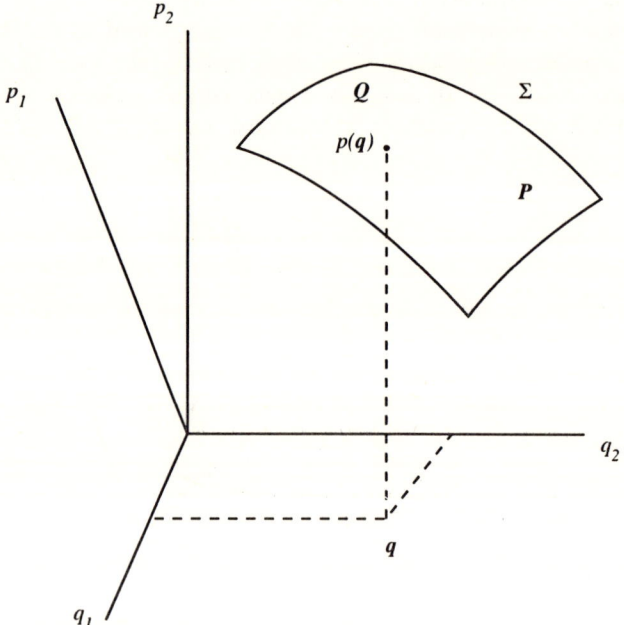

Figure 7.3: Lagrangian surface Σ

---

[2] The discussion here largely follows Berry (1983) and Batterman (1995).
[3] For the technical definition, of which we need not be concerned here, see Littlejohn (1992) and Arnold (1989).

mentum coordinates $\mathbf{P} = (P_1, P_2)$. The full set $(\mathbf{Q}, \mathbf{P})$ might, for instance, be a set of angle/action coordinates. We can now think of $(\mathbf{Q}, \mathbf{P})$ as providing a different coordinatization of the entire phase space that is connected to the original $(\mathbf{q}, \mathbf{p})$ by a canonical transformation provided by a generating function $S(\mathbf{q}, \mathbf{P})$—a function of both old and new canonical variables:

$$\mathbf{p} = \nabla_{\mathbf{q}} S(\mathbf{q}, \mathbf{P}) \tag{7.1}$$
$$\mathbf{q} = \nabla_{\mathbf{p}} S(\mathbf{q}, \mathbf{P}). \tag{7.2}$$

Schematically, the generating function operates as follows:

$$(\mathbf{q}, \mathbf{p}) \leftarrow S(\mathbf{q}, \mathbf{P}) \rightarrow (\mathbf{Q}, \mathbf{P}).$$

The Lagrangian surface is, then, the set of points in phase space $(\mathbf{q}, \mathbf{p}) = (\mathbf{q}, \nabla_{\mathbf{q}} S(\mathbf{q}, \mathbf{P}))$. The generating function plays the role in mechanics played by the optical distance function $\phi(\mathbf{r}, \mathbf{R})$ in geometrical optics.

Now the idea is to associate a (semiclassical) wavefunction $\psi(\mathbf{q})$ with this Lagrangian surface in such a way that the probability density (the "intensity") is proportional to the density of points in the classical *coordinate space*—the (q_1, q_2) plane in figure 7.3. The density is determined by "projecting down"— taking $|d\mathbf{Q}/d\mathbf{q}|$—from Σ onto **q**-space, assuming points are uniformly distributed in $\mathbf{Q}$. We introduce the *ansatz* for the wavefunction $\psi(\mathbf{q})$:

$$\psi(\mathbf{q}) = a(\mathbf{q}) e^{ib(\mathbf{q})}, \tag{7.3}$$

with amplitude $a(\mathbf{q})$ and phase $b(\mathbf{q})$. Given this, the requirement on probability density requires (using [7.2]):

$$a^2(\mathbf{q}) = K \left| \det \left\{ \frac{\partial^2 S(\mathbf{q}, \mathbf{P})}{\partial q_i \partial P_j} \right\} \right|, \tag{7.4}$$

where K is a constant.

The phase $b(\mathbf{q})$ is determined by appeal to the de Broglie relation, $\vec{p} = \hbar \vec{k}$, which relates the momentum $p(\mathbf{q})$ to the wavevector of a locally plane wave. This introduces Planck's constant and yields, after some straightforward manipulation involving equation (7.1), the phase:

$$b(\mathbf{q}) = \frac{S(\mathbf{q}, \mathbf{P})}{\hbar}. \tag{7.5}$$

Putting all of this together finally yields the wavefunction:

$$\psi(\mathbf{q}) = K \left| \det \left\{ \frac{\partial^2 S(\mathbf{q}, \mathbf{P})}{\partial q_i \partial P_j} \right\} \right|^{1/2} e^{\frac{i}{\hbar} S(\mathbf{q}, \mathbf{P})}. \tag{7.6}$$

Note the similarity between this formula and a single term of the interfering ray sum (6.14). In particular, note that $\hbar$ appears in the denominator of the exponent. This indicates that (as expected) the limit $\hbar \to 0$ of (7.6) is singular.

I mentioned in my discussion of the optics case that the interfering ray sum can also be arrived at through asymptotic analysis of the Helmholtz equation. The fact that this "top down" derivation agrees with the "bottom up" construction just performed serves to validate the association of wavefunctions with classical Lagrangian surfaces. It is worthwhile in the current context to provide the beginnings of the top down analysis. It is the so-called WKB method familiar to practitioners of quantum mechanics. The typical attitude of both physicists and philosophers toward this method is that it is simply an approximation scheme that is useful when the Schrödinger equation cannot be solved exactly. My view, on the other hand, is that the WKB method serves as the primary tool for understanding the asymptotic domain between classical and quantum mechanics. It is much more than a pragmatically useful method for solving computationally difficult problems.

7.2 The WKB Method

In this section I will show how, starting with the Schrödinger equation, we can arrive at a top down asymptotic expression for the wavefunction of the form (7.6). We will then see that this solution exhibits the singular features—particularly divergence at caustics—that we found when considering the corresponding situation in the optics case.

Let's begin with the one dimensional Schrödinger equation:

$$i\hbar \frac{\partial \psi}{\partial t} = -\frac{\hbar^2}{2m}\frac{\partial \psi}{\partial x} + (V(x) - E)\psi. \tag{7.7}$$

Now write the wavefunction $\psi(x)$ in the following form:

$$\psi(x) = e^{\frac{i}{\hbar}S(x)}, \tag{7.8}$$

where S is, in general, a complex function of x. Next, we substitute (7.8) into (7.7) and solve for S.

$$i\hbar \frac{\partial e^{\frac{i}{\hbar}S(x)}}{\partial t} = -\frac{\hbar^2}{2m}\frac{\partial e^{\frac{i}{\hbar}S(x)}}{\partial x} + (V(x) - E)e^{\frac{i}{\hbar}S(x)}. \tag{7.9}$$

Carrying out the differentiations yields

$$0 = \left\{\frac{1}{2m}\left(\left(\frac{\partial S}{\partial x}\right)^2 - i\hbar \frac{\partial^2 S}{\partial x^2}\right) + (V(x) - E)\right\} e^{\frac{i}{\hbar}S(x)}$$

hence, $$0 = \frac{1}{2m}\left(\frac{\partial S}{\partial x}\right)^2 + (V(x) - E) - \frac{i\hbar}{2m}\frac{\partial^2 S}{\partial x^2}. \tag{7.10}$$

This equation is exact. Next we expand the function S in powers of $\hbar$:

$$S(x) = S_0(x) + \hbar S_1(x) + \frac{\hbar^2}{2}S_2(x) + \cdots. \tag{7.11}$$

Intertheoretic Relations—Mechanics

It turns out that this series is *asymptotic*. It has the characteristic property of asymptotic series that if one truncates it after a few terms and then considers the result for small values of the parameter $\hbar$, one will have a good approximation to the actual value, *even though the series, considered in its entirety, is divergent.*[4] In other words, as we include more terms in the series, the numerical result will eventually diverge without bound from the true value of S.

If we now insert (7.11) into (7.10) and group terms according to the power of $\hbar$ they multiply, the result, retaining terms only up to and including $\hbar^2$, is:

$$0 = \frac{1}{2m}\left(\frac{\partial S_0}{\partial x}\right)^2 + (V(x) - E) + \frac{\hbar}{2m}\left(\frac{\partial S_0}{\partial x}\frac{\partial S_1}{\partial x} - \frac{i}{2}\frac{\partial^2 S_0}{\partial x^2}\right)$$
$$+ \frac{\hbar^2}{2m}\left(\left(\frac{\partial S_1}{\partial x}\right)^2 + \frac{\partial S_0}{\partial x}\frac{\partial S_2}{\partial x} - i\frac{\partial^2 S_1}{\partial x^2}\right).$$

This equality can hold only if the coefficient of each power of $\hbar$ is separately equal to zero. Thus, we have

$$\frac{\partial S_0}{\partial x} = \pm\sqrt{2m(E - V(x))} \tag{7.12}$$

$$S_0 = \pm\int_{x_0}^{x}\sqrt{2m(E - V(x))}dx \tag{7.13}$$

$$S_1 = \frac{i}{2}\ln\frac{\partial S_0}{\partial x}. \tag{7.14}$$

Finally, by inserting these values for S_0 and S_1 into the expansion of $S(x)$ in the wavefunction (7.8), we get the following WKB expression for ψ (with A and B arbitrary constants):

$$\psi(x) = \frac{A}{\sqrt[4]{2m(E - V(x))}}e^{\frac{i}{\hbar}\int_{x_0}^{x}\sqrt{2m(E-V(x))}dx} \tag{7.15}$$

$$+ \frac{B}{\sqrt[4]{2m(E - V(x))}}e^{-\frac{i}{\hbar}\int_{x_0}^{x}\sqrt{2m(E-V(x))}dx}. \tag{7.16}$$

This is the first-order WKB asymptotic approximation. It is the superposition of a wave traveling in the positive x direction (positive exponential) and a wave traveling in the negative x direction (negative exponential). It is formally similar to the bottom up construction provided by equation (7.6). A little further analysis will show how similar it is to the interfering ray sum (6.14).

Note that when $V(x) = E$, the denominators in both terms are infinite. Classically, this identity characterizes a turning point—a point where a classical particle would have zero momentum and its direction of motion would change. In the classical theory the only allowed region for a particle is that characterized

---

[4]For a discussion of asymptotic series in this context, see Batterman (1997).

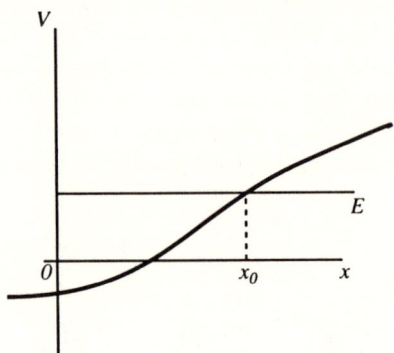

Figure 7.4: Potential barrier

by its kinetic energy being greater than the potential energy. One should keep figure 7.4 in mind here. When $x \leq x_0$, $V(x) \leq E$ and a classical particle coming from the left must turn around at x_0. The WKB solution to the Schrödinger equation, however, allows the wavefunction to be valid even in regions that are classically forbidden—when $V > E$.

As noted, at the classical turning points the amplitude of the wavefunction blows up. This is completely analogous to the singular behavior of the interfering ray sum at the caustic. In fact, classical turning points in one dimension *are* caustics. Furthermore, it appears that we can provide an interpretation for the phase $S(X)$ in analogy with the optical distance function ϕ appearing in the phase of the interfering ray sum. Constant values for S are classical surfaces in phase space similar to geometrical wavefronts, and normals to these surfaces are classical trajectories in analogy with the rays of geometrical optics. The semiclassical (small $\hbar$) asymptotic agreement between the top down analysis provided here and the bottom up construction of section 7.1 provides a justification for our association of wavefunctions with classical surfaces in this asymptotic domain. We can draw the same conclusions about this quantum/classical "no man's land" that we were able to draw about the shortwave catastrophe optics domain existing between the wave and ray theories of light. Just as in the case of light there are structures—features that appear in the semiclassical limit—whose understanding requires reference to classical structures and are not fully explicable (without asymptotic analysis) from the perspective of the quantum theory alone.

Much research in recent years has been devoted to studying these emergent structures and their patterns. I'll have a bit more to say about some of this later. However, most of this research has tended to focus on the fact (mostly ignored by my discussion until now) that quantum mechanics is a dynamical theory. We need to discuss this aspect.

The Schrödinger equation tells us how the wavefunction evolves over time.

Intertheoretic Relations—Mechanics

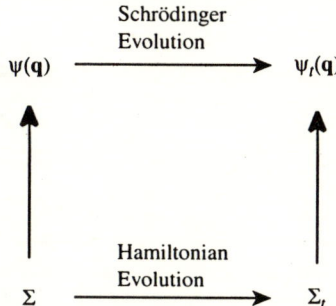

Figure 7.5: Invariance of the semiclassical construction over time

One question we might reasonably consider is whether the classical evolution of Lagrangian surfaces under Hamilton's equations is mirrored by the Schrödinger evolution of the constructed wavefunctions in the semiclassical limit. Figuratively, we are asking whether the diagram of figure 7.5 is satisfied.

The answer is that for a restricted class of Lagrangian surfaces, the invariance of the semiclassical construction over time does hold. This is the class of Lagrangian surfaces that are invariant under Hamiltonian evolution. They have the topology of N-dimensional tori (or doughnuts) in the $2N$-dimensional phase. Classical dynamical systems whose trajectories are confined to invariant tori are known as *integrable* systems. They exhibit what is called "regular" motion—that is, periodic or multiply periodic motion. The simplest example is the simple harmonic oscillator whose phase space trajectory just *is* the torus in one dimension. Recall figure 7.1. The semiclassical mechanics of systems whose "underlying" classical motions are regular has been around for a long time. In essence it is a refinement of the "old" quantum theory of Bohr, Sommerfeld, and Born, et al. Sometimes this theory goes by the name "EBK" theory, after Einstein, Brillouin, and Keller. However, Berry's term, "Torus Quantization" is much more descriptive.[5]

On the other hand, most classical evolutions are not confined to invariant tori. Those systems exhibiting irregular and even fully chaotic motions are far more common in classical mechanics than the typical textbook examples of integrable or regular systems. This fact causes problems for torus quantization and, as a result, for the invariance of the semiclassical association of wavefunctions with Lagrangian surfaces.

Under noninvariant evolution the Lagrangian surface of figure 7.3 may develop folds and other more complicated morphologies. For instance, in figure 7.6 the surface has evolved so as to yield, upon projection to the coordinate space, (q_1, q_2), a fold caustic. Each point inside the fold results from two distinct points (e.g., points 1 and 2) on the surface. Once again this demonstrates how the

---

[5]See Berry (1983), Batterman (1991).

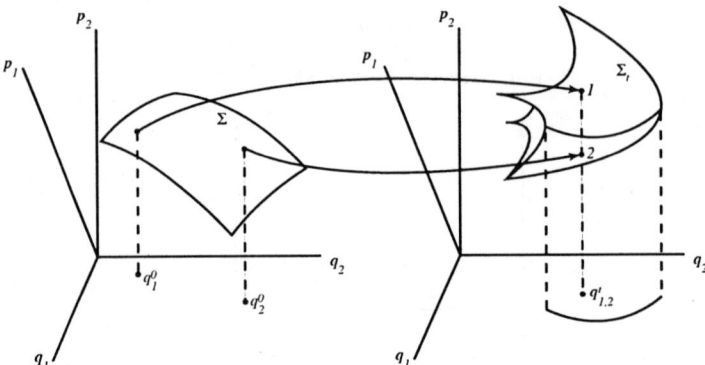

Figure 7.6: The folding of a Lagrangian surface under evolution

caustic organizes the multivaluedness of the "trajectories" in coordinate space. One might also have evolutions leading to cusp caustics, as the surface of figure 7.2 shows, as well as evolutions leading to other more complicated morphologies, some of which are classified by catastrophe theory.

As long as these foldings are relatively far apart, it is possible *at any given time* to use the recipe for associating (semi)classical wavefunctions with classical surfaces. The problem is that, when the folds become too close, then one will no longer be able to deal with the caustic singularities using Maslov's methods.[6] Neither will the WKB approximation be applicable. But exactly when do the foldings of the surface "become too close" for these methods to apply? In two dimensions there is a simple criterion for deciding this.

Suppose that an initial surface Σ_0 (the curve in figure 7.7) evolves over time Δt into the surface Σ_t. Suppose, further, that the points labeled 1, 2, and 3 (in figure 7.7 [a]) evolve to have the same q-value in time Δt. Then the constructed wavefunction associated with the surface Σ_t will be expressed as the superposition of contributions of the form (7.6) for each point above q_t in figure 7.7 (b). However, this construction will fail if the phase space area, A^{-t}, enclosed between the line connecting the pre-images of these points (e.g., points 1 and 2) *is of order* $\hbar$. This is what it means for the folds to be "too close." As long as the folds remain fairly isolated, it will be possible to use semiclassical formulas analogous to the diffraction catastrophe integrals (such as equation [6.19]) to construct good approximate wavefunctions. But if the Hamiltonian evolution is irregular or chaotic, then in the limit as $t \to \infty$, the catastrophes will no longer be isolable. The Lagrangian surface will become so convoluted that there will be ubiquitous clustering of caustics on scales smaller than order $\hbar$. In that case, the method of the diffraction catastrophe integrals itself will break down. This means that there is a clash between two fundamental

---

[6]Recall the brief discussion of Maslov's methods in section 6.3.

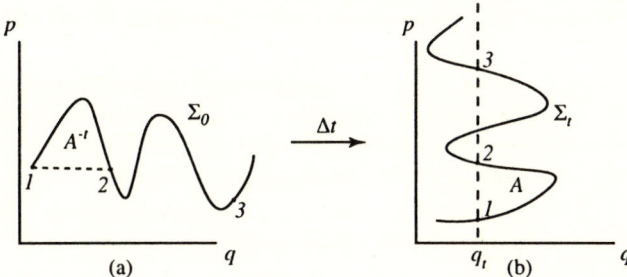

Figure 7.7: "Area" criterion for validity of constructed wavefunction

asymptotic limits—the semiclassical limit ($\hbar \to 0$), and the long-time limit ($t \to \infty$).

In fact, if the classical evolution is chaotic, then these limits fail to commute. That is, while for fixed t after some initial time t_0 it will be possible to keep the folds sufficiently far apart by letting $\hbar$ get very small, one can't do this if the $t \to \infty$ limit gets taken first. The caustics will cluster on too fine a scale. Since classical chaos is a $t \to \infty$ property, it is apparently not possible to recover that chaotic behavior in the $\hbar \to 0$ limit of quantum mechanics. This is the so-called problem of quantum chaos.

7.3 Semiclassical "Emergents"

For my purposes here, the problem of quantum chaos provides further evidence for a failure of limiting reductive relations between quantum and classical mechanics. Much research of late has been devoted to studying various aspects of the asymptotic clash between these two limits and certain features and structures that emerge in these asymptotic domains. The asymptotic "no man's land" is much more complex and interesting than Bohr and many others had thought. It should be clear that a straightforward interpretation of the correspondence principle simply does not appear to be available.

There are many aspects of the semiclassical limit of quantum mechanics that cannot be explained purely in quantum mechanical terms, though they are in some sense quantum mechanical. Just as the case of optics discussed in chapter 6, the fundamental theory is, in certain instances, explanatorily deficient. For instance, there are differences in the morphologies of semiclassical wavefunctions that depend *essentially* on the nature of the "underlying" classical motion—whether the classical motion is regular or chaotic. When the motion is integrable or regular, then, as I've mentioned, torus quantization applies and one finds fairly simple formulas for constructing wavefunctions. When the classical motion is chaotic, the tori break down; nevertheless, classical periodic trajectories (now isolated—that is, not appearing in families—yet densely distributed in a sea of nonperiodic trajectories) still play an essential role in fixing the form of the wavefunction's probability density. This phenomenon is called "scarring" and the "rule" for quantization is sometimes called the "Gutzwiller trace formula."[7]

Another feature among many that are emergent in the semiclassical domain is the statistical behavior of the spacings of the energy levels for the quantum systems. If we consider a bound quantum system—one with a discrete spectrum of energy levels—we can ask how these levels are distributed in the semiclassical limit. As $\hbar \to 0$, and the classical limit $\hbar = 0$ is approached, we should rightly expect that the energy levels get closer and closer together. In the classical theory, after all, the energy is distributed along a continuum. The mean spacing between levels when $\hbar$ is small is proportional to $\hbar^N$, where N is the number of degrees of freedom of the system. If we then "look" at the levels with a microscope with magnification $\hbar^{-N}$ (in other words, we rescale the level spacings in proportion to the "size" of $\hbar$), we will be looking at a set of levels with a unit mean density that remains constant as $\hbar \to 0$. Now we can ask about how the distribution of levels fluctuates about this mean. Here is what Berry has to say about the answer to this question:

> The answer is remarkable: apart from trivial exceptions, the fluctuations are *universal*, that is, independent of the details of the system and dependent only on whether the orbits of its classical counterpart are regular or chaotic. Paradoxically, the spectral fluctuations are those of a sequence of random numbers (Poisson distribution) when the classical motion is regular, and are more regularly distributed (exhibiting the level repulsion characteristic of the eigenvalues of random matrices) when the classical motion is chaotic. We are beginning to understand this quantum universality in terms of semiclassical asymptotics: it arises from a similar universality in the distribution of long-period classical orbits. (Berry, 1994, p. 604)

It is indeed remarkable how these quantum mechanical features require reference to classical properties for their full explanation. Once again, these features are all contained in the Schrödinger equation—at least in the asymptotics of its combined long-time and semiclassical limits—yet, their interpretation requires reference to classical mechanics.

7.4 Conclusion

In this chapter and in chapter 6 I have discussed examples of complex phenomena that apparently inhabit the asymptotic borderland between theories related by singular limits. I have shown how one can explore the nature of these phenomena in a sense from two "directions"—from the bottom up and from the top

---

[7]For a nice, relatively nontechnical survey article about some of these semiclassical emergent features, see Heller and Tomsovic (1993). See also my discussions Batterman (1991, 1993, 1995) and the book by Gutzwiller (1990).

down. The top down approach, which begins with the fundamental equation of the finer theory and investigates its solutions in an appropriate asymptotic limit, has been the main focus of this chapter. In contrast, the bottom up approach was the main consideration of chapter 6. We see that in the context of time evolution—that is, that when we investigate the long-time limit ($t \to \infty$) in combination with the semiclassical ($\hbar \to 0$) correspondence limit—things become much more complex and interesting.

These examples serve to indicate that (1) the neo-Nagelian reductive relations are woefully inadequate and that (2) the failure of the physicists' reductive schema (6.1), $\lim_{\epsilon \to 0} T_f = T_c$, leads one to investigate what can only be called "new physics."

My discussion of these borderland phenomena and their explanations should also lead one to question what has become a fairly common sentiment among physicists and philosophers. It seems that most investigators maintain a reductionist and even eliminativist attitude toward classical physics given the many successes of the quantum theory. Of course, it is just this attitude that I have been at pains to question. Nevertheless, I have frequently heard the line that one really need not pay any attention to classical mechanics. A typical expression of this attitude might be the following:

> Everything is ultimately quantum mechanical. The standard procedure of beginning with a classical system and then quantizing it according to well-established heuristic rules is simply old-fashioned. Ideally, we should simply treat all systems as quantum mechanical and our use of classical methods is purely pragmatic—simply a matter of convenience and a function of our deep, yet flawed, need to think classically. Once the revolution is complete and a new generation of physicists arrives, such classical thinking should really be jettisoned altogether.

I hope that the discussion of the last two chapters shows how misguided this way of thinking really is. Given that the understanding of various observed and observable universal behaviors "contained in" the finer theories (wave optics and quantum mechanics) requires reference to structures present only in the coarser theories to which they are related by asymptotic limits, it is indeed difficult to see how this attitude can coherently be maintained.

8

Emergence

I think it is fair to say that recently the topic of emergence has received most of its attention in the context of philosophy of mind. While there may be things distinctive about the emergence of mental phenomena, these discussions have also by and large tried to encompass the emergence of other special science properties as well.[1] My discussion of philosophical views about reduction also focused on issues about special science properties generally. It is, therefore, worthwhile for me to begin this chapter with a discussion of emergence in this context as well.

Let me say, however, that most of the discussion in this chapter will concern an example of emergence that is not from the special sciences. I will argue that the universal features of the rainbow—the way the fringe spacings and intensities scale in the neighborhood of a caustic—are emergent properties. While certain aspects of these phenomena satisfy certain features of the philosophical conception of emergence outlined in section 2.4, there are other features of that conception that are wholly inapplicable in this context. In particular, I will argue that the part/whole aspect of the "received view" of emergence is completely absent in this example, though issues about irreducibility, unexplainability, and unpredictability remain salient. Furthermore, I will argue that the mereological aspects of the usual view of emergence are relatively *unimportant* vis-à-vis emergence even in those contexts—even for those properties—for which they appear to play an essential role. Instead, what is important, and often missed, is the *asymptotic nature* of the relationship between the emergent properties and the lower level theory characterizing the components.

Section 8.1 reviews what I'll call the "received" view of emergence. In section 8.2 I will continue the discussion of the rainbow example. I will consider the extent to which the universal scaling properties of light in the neighborhood of a caustic can be considered emergent properties. Next, in section 8.3 I will formulate a new sense of emergence—one that captures and accounts for several

---

[1] Kim, for instance, thinks that if any property can legitimately be understood as emergent, it will be a phenomenal feel or "quale." On his view Kim (1999), as we will see later, most other "upper level" special science properties will fail to be emergent.

of the key features of the "received" view. Finally, in section 8.4 I will try to make sense, on this new view, of the idea that emergents possess novel causal powers. Actually, I will argue there that this novelty is best understood not in terms of causation but rather in terms of the existence of new theories that play novel explanatory roles.

8.1 Emergence and the Philosophy of Mind

Let's begin by reviewing and expanding the discussion of section 2.4. I noted there that there are several assumptions underlying what we can take to be the "received" view about emergence in the literature today.[2] The first assumption is that the world divides somehow into levels. Thus, an important difference between a special science, such as psychology, and more "basic" underlying sciences, such as neurophysiology and physics, is the fact that the things talked about in psychology exist in some important sense at a higher (= less fundamental) level.

The second assumption is that this conception of levels is to be understood in mereological terms. In other words, properties of wholes existing at some level are constituted by the basic parts of that whole and their various features. For example, Kim expects most emergentists to accept what he calls "mereological supervenience."[3]

- *Mereological Supervenience*: Systems with an identical total microstructural property have all other [intrinsic] properties in common. Equivalently, all properties of a physical system supervene on, or are determined by, its total microstructural property. (Kim, 1999, p. 7)

Kim notes further that those who hold that there exist emergent properties typically need to try to distinguish them from other upper level properties that are merely *resultant* (1999, pp. 7-8). This difference has traditionally been founded on epistemological considerations of predictability and explainability. In particular, emergent properties are said to be *novel* at least in the sense that they are neither explainable nor predictable from the theory of the lower level constituents.

More recently, perhaps, issues about reducibility have also played a role in distinguishing those properties that are emergent at a higher level from those

---

[2] There is no unanimity in this literature. Nevertheless, I think Kim is right to think that the four (or five) tenets he identifies are quite prevalent in contemporary discussions.

[3] A notable exception to this is Paul Humphreys (1997). He suggests that, to meet various objections against the possibility of emergent properties, one actually needs to give up on supervenience. For him, emergent properties are formed by an ontological operation in which two (or more) lower level property instances fuse to form a property instance at the next higher level. As a result of this fusion, the lower level property instances no longer exist; and hence, there is no supervenience base upon which the emergent properties can supervene. Note, however, that on this view, the fusion operation is mereological. The emergent properties are formed by a real coming together of lower level parts and their properties, even if once the operation has been effected, such parts no longer exist.

Emergence

that are merely resultant and pose no (or little) interpretive problems. Most authors assume (and Kim is no exception) that reductions are essentially explanatory. We have seen, however, that there are good reasons to think that reduction and explanation can part company. Thus, there are no good reasons to maintain that reduction (in all of its guises) need be essentially epistemological.[4]

As I noted in section 2.4, Kim tries to capture contemporary thinking about emergent properties in the following five tenets:

1. *Emergence of complex higher-level entities*: Systems with a higher-level of complexity emerge from the coming together of lower-level entities in new structural configurations.

2. *Emergence of higher-level properties*: All properties of higher-level entities arise out of the properties and relations that characterize their constituent parts. Some properties of these higher, complex systems are "emergent," and the rest merely "resultant."

3. *The unpredictability of emergent properties*: Emergent properties are not predictable from exhaustive information concerning their "basal conditions." In contrast, resultant properties are predictable from lower-level information.

4. *The unexplainability/irreducibility of emergent properties*: Emergent properties, unlike those that are merely resultant, are neither explainable nor reducible in terms of their basal conditions.

5. *The causal efficacy of the emergents*: Emergent properties have causal powers of their own—novel causal powers irreducible to the causal powers of their basal constituents (Kim, 1999, pp. 20–22).

It is clear that the first two tenets focus on the part/whole nature of the emergent properties, while the last three are concerned with trying to capture the sense in which the emergents are novel. In the next section, I will consider again the example of the rainbow. I hold that the patterns of fringe spacings and intensities that exist in the asymptotic limit near the caustic are genuinely emergent properties. We will see, however, that there are no part/whole relations in play here. I will also give an account of the different senses in which these patterns are novel. This will enable me to refine the conception of emergence in several important ways.

8.2 The Rainbow Revisited: An Example of Emergence?

In this section, I will *assume* that we do want to think of the patterns of fringe spacings as emergent phenomena and examine the extent to which these phe-

---

[4]See Antony (1999) for a similar point of view explicitly in the context of a discussion of emergence.

nomena satisfy the tenets of contemporary emergentist thinking noted above in section 8.1. The discussion of section 8.3 will show that this is a legitimate assumption.

To begin, let us recall Berry's characterization of the scaling laws for the fringe spacings and the intensities of the wavelike aspects that decorate caustics in the shortwave limit.

> [The] "diffraction catastrophes" have intricate and beautiful structures, and constitute a hierarchy of nonanalyticities, of emergent phenomena par excellence. The patterns inhabit the borderland between the wave and ray theories, because when λ is zero the fringes are too small to see, whereas when λ is too large the overall structure of the pattern cannot be discerned: they are *wave* fringes decorating *ray* singularities. (Berry, 1994, pp. 602–603)

As we've seen, an important assumption underlying the "received" characterization of emergence is that the world be construed as consisting of a hierarchy of levels. The things talked about by theories at upper levels are to be related mereologically to things talked about by lower level, typically *more fundamental*, theories. In the context of my discussion of the physicists' sense of intertheoretic reduction, we would typically take the coarser theory to be a higher level theory related to the finer, more fundamental theory by the schema (6.1). For instance, we have the coarser theory of Newtonian space and time related by this schema to the finer, more basic, theory of special relativity. Even in the cases (the most interesting cases) for which the reductive schema fails, such as the relationship between the ray and wave theories of light or that between classical and quantum mechanics, it does seem to make sense to consider the second member of these pairs to be more fundamental in an important sense. After all, the "coarser" members of the pair are, strictly speaking, false, and they have been superseded by the second, "finer" members.

Despite the difference between the more and less fundamental nature of the theories in these pairs, it is difficult to hold that this difference maps nicely onto the conception of levels that is employed in most philosophical discussions about emergence. For instance, it is hard to see any sense in which the properties of the "entities" or systems of classical mechanics are composed of parts and properties of quantum mechanical systems. Similarly, it doesn't make sense to speak of the rays that are the basic entities of the ray theory as having properties that arise through the coming together of wave theoretic components and their properties and relations. If we are trying to understand and explain certain features of the rainbow, for instance, *both theories seem to be operating at the same level*. "Level" talk here seems completely inappropriate. Instead, it makes better sense to speak of a distinction between *scales*. Ray theoretic talk becomes more appropriate on scales in which the wavelength is *small*. Quantum mechanical effects dominate at scales in which Planck's constant is *large* in comparison with "classical" quantities having the same dimension. While we surely do want to think of quantum mechanics and the wave theory as more fundamental

(= basic?), it is surely not the case that that notion is captured by the hierarchy of levels model where that model depends upon the coming together of wholes from parts as one moves up the hierarchy.

But someone might respond to this as follows. Why not think that the caustic structure of the ray theory is some kind of component property of the phenomena we observe in the rainbow? After all, I noted that often investigators speak of the "wave flesh decorating the ray theoretic bones." So we might take these bones to be components out of which the wave phenomena is built. Furthermore, recall the discussion of the interfering ray sum equation (6.14). There I attempted to construct a wavefunction through a summation—through some kind of aggregation—of rays. So, even if it might be difficult to think of the wave and ray theories as fitting in a hierarchy of *levels*, we may still be able to maintain a mereological connection between various entities and properties described by the two theories. And so, at least some aspects—in particular, the mereological supervenience that is often assumed in the contemporary emergentist position—would apply to the current example of emergent properties.

There are two main difficulties with this response. On the one hand, if we take the more fundamental wave theory to be the *base* theory, then this line of reasoning gets things backwards. Rays are *not* objects of the base theory, waves are. Clearly this is related to the different ways philosophers and physicists think about reduction. Philosophers would surely (and correctly) take the wave theory to be more fundamental and look for a reduction of the ray theory to the wave theory. Physicists, as we've seen, would have it that the wave theory reduces to the ray theory in the limit.[5] They, too, would correctly think of the wave theory as being more fundamental, though not in the sense of parts being more "basic" than wholes. But the tenets of emergentism, particularly tenets 1 and 2, assume the philosophical picture of this aspect of intertheoretic relations. The "inversion" that seems to be required to map the rainbow example onto the philosophical characterization of emergence is just one more indication that the traditional philosophical thinking about intertheoretic—reductive—relations is wrong-headed.

The second reason why this response—thinking of the wave phenomena of interest as being due to an aggregation of ray theoretic components and their properties—is misguided has to do with the nature of the construction (equation [6.14]) itself. Constructing this asymptotic approximation from ray theoretic components required that we endow each ray with a phase,

$$k\phi_\mu(\mathbf{R}, z),$$

with $k = 2\pi/\lambda$. In other words, we had to associate a wave theoretic feature with each ray in the construction. The construction is helpful because it allows one to see fairly clearly what the relationship is between the ray theory and the wave theory in the asymptotic shortwave limit for waves relatively far from

---

[5] As schema (6.1) fails because of the nonanalytic nature of the limit, the reduction fails. Nevertheless, many physicists, including Berry, still talk about reduction. As noted in the last chapter, I prefer to speak of intertheoretic relations.

ray theoretic caustics. It provides a fairly intuitive picture of the asymptotic domain of interest. But the *ultimate justification* for the (limited) applicability of this construction is the fact that *asymptotic analysis starting with the wave equation*—the Helmholtz equation (6.12)—*yields the same interfering ray sum in the stationary phase approximation.*

So, the idea that one can take the emergent wave theoretic features of the rainbow as "arising from" a "coming together" (see tenets 1 and 2) of ray theoretic components and their properties—as the tenets of emergentism would have it—doesn't work. Wave theoretic features must be built in from the beginning. The part/whole features characteristic of the "received" emergentist position simply fail to apply to the case of emergence under consideration. The "naked emergentist intuition"—namely, that "an emergent property of a whole somehow 'transcends' the properties of the parts"[6]—just fails to apply to the emergent phenomena I have been considering.

Now let me begin to consider the novelty aspects of the emergentist position. Tenets 3 and 4 refer to the unpredictability and unexplainability/irreducibility of the emergent properties in terms of the *basal conditions*. I must now discuss the possibility of explaining and predicting the emergent rainbow phenomena.

Kim understands "basal conditions" to be "the lower-level conditions out of which [emergent properties or phenomena] emerge" (1999, p. 6). I have already expressed worries about how to understand the hierarchical distinction between basic and nonbasic in this context where it appears the traditional conception of a level hierarchy fails to apply. Nevertheless, given the view that the wave theory is more fundamental, it seems reasonable to understand "basal conditions" as referring to aspects of the wave theory in this context. So the question becomes: Are the emergent aspects of the rainbow—those features captured by the scaling law (6.21) for fringe spacings and intensities—explainable and predictable in terms of the wave theory?

Now, if by "explanation" one has in mind one of the dominant contemporary conceptions—the D-N model or some causal-mechanical model—then I think that the answer is a clear "no." The discussion in chapter 6 about the nature of the diffraction catastrophe scaling laws, together with the discussion of asymptotic explanation in chapter 4, makes this clear. These emergent phenomena are not derivable in any straightforward sense from the underlying wave theory. They are not, as it were, from-first-principle solutions to the wave equation (6.12). They are deeply encoded in that equation but are apparent only through its asymptotic analysis.

On the other hand, the "theory" of that asymptotic domain—catastrophe optics—does provide satisfactory accounts of the phenomena of interest. So the phenomena are not inexplicable or brute *tout court*. One has, in a well-defined sense, an explanation that is "grounded in" the fundamental wave theory; but this kind of *asymptotic* explanation is, as we've seen in the discussion in chapter 4, distinct from the usual types of explanation talked about in the philosophical literature. The phenomena are not explainable through *derivation*—that

---

[6]See Teller (1992, p.139).

is, through straightforward solutions to the differential equation—from the fundamental wave theory alone.

Recall the discussion (chapter 5) of the explanation of universality in philosophy of mind in particular, and the special sciences in general. There, I was concerned in part to maintain a robust physicalism while at the same time respecting the multiple realizability argument against reduction. The solution was to see that multiple realizability is just an instance of universality. Also, explanations of universal (multiply realized) behavior proceed by finding principled reasons *in the physical theory of the realizers* for the irrelevancy of certain details that otherwise are quite important for answering certain explanatory questions. It is in this sense that the explanations are "grounded in" the physical theory yet are obviously quite different from the usual understanding of explanation by derivation from first principles.

Are the phenomena predictable from the wave theory? Again, I think that the answer here must be negative. If by "predicting the presence of phenomenon P," one means (as philosophers typically do) "it can be shown that P is an exact solution to the fundamental governing equation of the appropriate theory, given all the details one might require as to initial and boundary conditions," then these emergent phenomena are not predictable. On the other hand, things are, perhaps, a bit more tricky than this. After all, it seems reasonable to think that part of what it means to say that the wave theory is fundamental is that the governing equation of the theory—the wave equation—"contains" the sum total of wave phenomena. But, as I argued at the end of chapter 6, the real issue is how to understand properly this notion of containment. This is analogous to understanding the distinction, just highlighted, between an explanation being "grounded in" the fundamental theory but not itself being an instance of derivation from that theory. As I've argued, I don't think that when properly understood, the sense of containment entails predictability in the usual sense. In fact, it must be understood as referring to asymptotic analysis.

Furthermore, if I'm right and there is a genuine, distinct, third theory (catastrophe optics) of the asymptotic borderland between the wave and ray theories—a theory that of necessity must make reference to both ray theoretic and wave theoretic structures in characterizing its "ontology"—then, since it is this ontology that we take to be emergent, those phenomena are not predictable from the wave theory. They are "contained" in that theory but aren't predictable from it.[7]

All of this is just to say that with respect to many phenomena of interest, the "base" theory is often explanatorily and predictively deficient. This has been a major theme throughout this book. From-first-principle accounts often obscure what is genuinely explanatory. Asymptotic reasoning in its many guises, remedies this situation by highlighting the "principal laws" and dominant features of the phenomena of interest.

Let us now turn to the question of the reducibility of the scaling patterns for

---

[7]The same, of course, goes for the emergent phenomena characterized by semiclassical mechanics.

the fringes and the intensities of the rainbow. Are they *reducible* to the wave theory? I take it that the arguments of chapter 6 have shown that they are not, at least if reduction is to be understood in either the neo-Nagelian, philosophers' sense or in the physicists' sense captured by schema (6.1). Nor are these patterns reducible on Kim's functional model of reduction. We have seen (section 5.3) that that model fails to account for the important—universal—nature of the phenomena under consideration. The phenomena are, after all, universal features of light in the shortwave limiting domain. We want to know whether such lawlike patterns of light in the neighborhood of caustics are reducible to lawlike relations in the wave theory. At best, it seems, we can say the following about the functional model of reduction as applied to this (and other cases). *If* the phenomena are functionalizable, we will have accounts of why individual instances of the patterns are displayed on given occasions. *But this, as should now be familiar, in no way provides a reductive account of the lawlike universal pattern itself.*

As we have seen, the ultimate reason for the failure of reducibility is the singular nature of the asymptotic relationship between the wave and ray theories. It is because of this that we have the predictive and explanatory failures discussed before. In an important sense, then, *the novelty of the emergent phenomena I am discussing is a direct result of the singular nature of the correspondence relations between the two theories.*

To sum up this discussion so far, we can say that the emergent rainbow phenomena—the nature of the fringe spacings near the caustic—have the following features:

- They are not reducible to either theory involved—neither to the more fundamental wave theory nor to the coarser ray theory.

- They are not explainable in terms of the more fundamental theory (though there are asymptotic explanations that are "grounded in" that theory).

- They are not predictable from the properties of the more fundamental theory (though they are in a well-defined sense asymptotically "contained in" that theory).

Each of these features maps to some extent onto the tenets that characterize the contemporary conception of emergence. However, the part/whole and level hierarchy aspects of emergent properties expressed in tenets 1 and 2 most surely do not hold for our candidate emergent property.

I have still not discussed tenet 5—the causal efficacy of the emergents. Surely the fact that on the standard conception of emergence the emergent properties are to have causal powers of their own is an important feature of their novelty. I will return to this later on in section 8.4.

8.3 A New Sense of Emergence

There is a sense of the term "emergence" that is not captured explicitly by the five tenets already listed. But it is a sense of the term that better fits the status of phenomena which inhabit borderlands between theories. According to the *Oxford English Dictionary*, "emergence" can be defined as "the process of coming forth, issuing from concealment, obscurity, or confinement." This explicitly reflects a process view of emergence that is not obviously present in the "received" philosophical conception I have been discussing. It is true that tenets 1 and 2 speak of emergents "arising out of" or resulting from the "coming together" of lower level parts and their properties. Nevertheless, these conceptions refer to combinatorial or aggregative operations and not to a kind of "continuous" process of issuing from concealment.

Of course, I do not intend much to hang on a dictionary definition of the term. However, the sort of limiting asymptotic relations that play essential roles in understanding and explaining the features of the rainbow are much more amenable to this sense of emergence as issuing from concealment, obscurity, or confinement. Mathematical asymptotics show exactly how the scaling laws come to dominate in the shortwave domain in which λ can be considered small. Asymptotic methods explain why we do not see the patterns when the wavelength is too big, and they enable us to understand the nature and behavior of the changing patterns as the limit $\lambda \to 0$ is approached.

On my view, the novelty of these emergent properties is, as I've said, a result of the singular nature of the limiting relationship between the finer and coarser theories that are relevant to the phenomenon of interest. It is therefore intimately tied to the failure of reduction as understood by the physicists' schema (6.1). I claim that this is true even in some cases where there is a natural level hierarchy defined in terms of parts and wholes. The thermodynamics of critical phenomena is just such a case. Let us reconsider this case with an eye toward demonstrating the essential singular asymptotic aspect of the emergent phenomena.

The two theories are classical thermodynamics of fluids and statistical mechanics. I have already mentioned reasons for doubting that anything like a straightforward Nagelian reduction of thermodynamics to statistical mechanics is possible. The reasons are complex and involve the fact that there are essential probabilistic elements involved in the reducing theory—statistical mechanics—that are completely absent in the reduced theory of thermodynamics. Perhaps some sort of neo-Nagelian approach is more appropriate. Recall that in cases of heterogeneous reduction, it is often the case that the reduced theory is corrected in the process of carrying out the "reduction." Thus, most investigators now think of thermodynamics as allowing for fluctuations. The second law, for example, becomes a statistical statement to the effect that entropy is overwhelmingly likely to be nondecreasing in an isolated system, though "antithermodynamic" fluctuations are possible. (See Sklar (1993) for an extended discussion.)

The connections between thermodynamics and statistical mechanics are very subtle. Nevertheless, it is possible to give derivations of suitably reinterpreted thermodynamic laws, in certain situations, from statistical physics. These derivations involve, essentially, the taking of asymptotic limits. In this context, the limit is called the "thermodynamic limit." As a relatively simple example, let's consider the "zeroth law" of thermodynamics. This is the assertion that there exists a property, *temperature*, that takes the same value for any two bodies in thermal contact. What is involved in "reducing" this law to statistical mechanics?

First of all, we need a way of modeling two bodies in thermal contact. We do this by treating an isolated classical system S (a system composed of molecules interacting via collisions) as consisting of two subsystems S_1 and S_2 that interact with one another across a boundary B. (See figure 8.1.) We need to assume that the interaction is sufficiently strong to allow energy transfer between the subsystems, but also that it is so weak that the interaction energy doesn't contribute significantly to the total energy of the system S. Next we assume that the dynamical motion of the composite system is ergodic and mixing. These are properties of the dynamics related to its instability. For my purposes here I don't need to go into their definitions. This assumption guarantees that we can use a particular probability distribution—the microcanonical distribution for the system S—to calculate the equilibrium values for various quantities measured on the subsystems S_1 and S_2.

The assumption that the interaction energy is sufficiently weak guarantees that to a first approximation we may treat the Hamiltonian for the composite system $H_S(\mathbf{q},\mathbf{p})$ to be the sum of the Hamiltonians of the components:

$$H_S(\mathbf{q},\mathbf{p}) = H_{S_1}(\mathbf{q_1},\mathbf{p_1}) + H_{S_2}(\mathbf{q_2},\mathbf{p_2}).$$

H_{S_1} is a function of the position and momentum coordinates of subsystem S_1, namely, $(\mathbf{q_1},\mathbf{p_1})$ and similarly for the Hamiltonian of subsystem S_2. Given that the composite system is isolated at constant energy E, the distribution in phase space for S at that energy is given by the delta function:

$$\delta[E - H_{S_1}(\mathbf{q_1},\mathbf{p_1}) - H_{S_2}(\mathbf{q_2},\mathbf{p_2})].$$

This distribution function enables us to determine the probability distribution for the energy E_1 of the first subsystem S_1:

$$p(E_1) = \frac{\Omega_1(E_1)\Omega_2(E - E_1)}{\Omega(E)}. \tag{8.1}$$

Here, Ω, Ω_1, and Ω_2 are the structure functions for the composite system and its components respectively. (See Khinchin (1949) for details and definitions.) The structure function Ω, for example, enables one to calculate the expected value for any measurable function over the energy surface for S. We can use equation (8.1) to find the *most probable value* for E_1.

I am now almost done. *If we are justified in neglecting fluctuations about this most probable value*, then we can show that the logarithmic derivative of the structure function is the *same* for the two subsystems in thermal contact.

Emergence

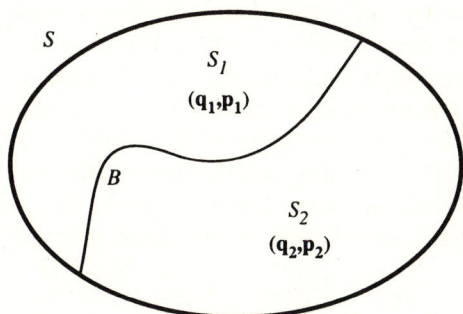

Figure 8.1: Systems in thermal contact

Thus, we can identify, or at least associate (via some functional relationship), this logarithmic derivative with the thermodynamic temperature. We have shown how to "derive" the zeroth law from statistical mechanics.

So many assumptions are involved in this "derivation" that it is likely not to seem terribly satisfying. From a philosophical perspective, one of the most troublesome aspects is the identification or association of nonstatistical thermodynamic functions with most probable values for certain functions of phase. Much work is required to interpret the sense of "most probable" here. The assumption that S is ergodic and mixing is also a very strong assumption, one that may be rarely justified in practice.[8] For my purposes here, the most important assumption is that we can neglect fluctuations about the most probable value. Much work has been done to justify this assumption. The received view is that we can do so, if we can show that the distribution of the energy E_1 is sharply peaked about the most probable value. One can show this for certain idealized[9] (and not so idealized)[10] systems by considering what is called the thermodynamic limit. This is the limit in which the number of particles, N, and the volume, V, go to infinity with the restriction that the ratio N/V—the density—remains constant. The point is that the thermodynamic limit can be shown to exist for certain systems (or for systems in certain regimes), and so the connection between thermodynamics and statistical physics is (relatively well) justified. What is crucial for us here is that the justification involves taking an infinite limit *as well as a demonstration that that limit exists*. The thermodynamic limit *does not* exist for systems undergoing phase transitions or for systems at their critical points.

The case we have been considering depends upon taking a limit involving the number of components at a lower level that make up the upper level composite

---

[8] Again, the best source for these discussions is Sklar (1993). See also Batterman (1998) for a discussion of ergodicity and justifying the use of the microcanonical distribution.

[9] See Khinchin (1949) for the case of an ideal gas with noninteracting components.

[10] See Mazur and van der Linden (1963) for an extension of Khinchin's results to systems with realistic interaction potentials between components.

system. It is, therefore, different from the rainbow situation in that parts and wholes do play a role. However, to make the connection between the theories—to explain and understand why thermodynamics works as well as it does—we needed to consider a limiting relationship as well. Asymptotic reasoning plays an essential role in establishing this connection. That is, in order to understand why the thermodynamic laws and relationships hold for a variety of physically distinct types of systems, we require a method that allows for a systematic abstraction from the details that distinguish the different systems—the realizers—from one another. The asymptotic reasoning involved here, given that it doesn't involve any physical singularities, is a type of reasoning analogous to what Barenblatt calls "asymptotics of the first kind."[11]

As it turns out, this relationship satisfies the reduction schema (6.1). Thus, it is analogous to the case of special relativity and the Newtonian theory of space and time. One can say that a reduction of the physicists' type obtains (at least partially) between thermodynamics and the underlying statistical mechanics of the components in the limit as $1/N \to 0$.

Given that schema (6.1) is satisfied, we must deny that the aggregation of the many components of the system (at least for systems not undergoing phase transitions or that are not at criticality) yields emergent phenomena. I take the failure of this reductive schema to be necessary for emergence. Not all aspects of wholes in thermodynamics are "greater than the sum of their parts." Those that are not constitute the "resultant" properties of the higher level phenomena. Despite this, in order to understand these resultants, asymptotic limits still play an essential role.

At criticality, however, things are very different. The limiting reduction fails completely. At criticality there are, as we've seen in section 4.1, long-range correlations. This is reflected by the fact that the correlation length actually diverges. As a result, fluctuations will no longer be negligible even in the thermodynamic limit as $N, V \to \infty$ with N/V constant. For such systems the "derivational" connections between thermodynamic laws and statistical mechanics will fail because the thermodynamic limit is singular.

The problem here is absolutely fundamental. Thermodynamics is a continuum theory—fluids are continuous blobs from the perspective of the upper level classical theory. The fact that they are made up of large numbers of interacting molecules is not considered. The reduction we have been considering must show that density fluctuations arising from intermolecular interactions occur only at a very small, microscopic, scale. On the other hand, at and near the critical point, such fluctuations exist at all scales up to the size of the fluid sample. This leads to the divergence (singularity) of the correlation length.

Berry puts the point as follows:

> Thus at criticality the continuum limit does not exist, corresponding to a new state of matter .... [T]he critical state is a singularity of thermodynamics, at which its smooth reduction to statistical mechanics breaks down; nevertheless, out of this singularity emerges a

[11] Recall the discussion in section 4.3.

large class of new "critical phenomena", which can be understood by careful study of the large-N asymptotics. (Berry, 1994, p. 600)

In fact, the approach to the critical point $(T \to T_c)$ is analogous to the approach to the caustic in our rainbow example. As these "structures" are approached, the relatively straightforward accounts (respectively, the argument just considered in the case of thermodynamics and statistical mechanics, and the interfering ray sum—equation [6.14]) completely break down. New methods, involving asymptotics of the *second kind* are required to understand and explain what is going on at and near these singularities.

The asymptotic investigation of the behavior of systems at and near these singularities lets us "observe" properties as they *issue from concealment and obscurity*. We are able to recognize and investigate emergent properties of these systems.

Let me try to sum up the discussion of this section. In the case of thermodynamical properties, it does make sense to think of upper level properties (as Kim would describe them) as "arising out of" the properties and relations characteristic of their constituent parts. Fluids and gases are composed of many interacting molecular components. However, this mereological, part/whole aspect is not the most important when it comes to emergent phenomena. Instead, what plays the most essential role is the fact that the thermodynamic limit is singular at the critical point. True, the divergence of the correlation length is due to a fact about parts and wholes—the correlation (at criticality) between widely separated components. This is, as I've said, responsible for the failure of the thermodynamic limit. But the explanation depends in a crucial way on the fact that the purported limiting connection between statistical mechanics and thermodynamics is singular. As the example of the rainbow shows, not all such singular behavior (in that case the singularity in intensity at the caustic according to the ray theory) results from some kind of blowup related to *parts* a composite *whole*. There are no parts in this case.

The proposed account of emergent properties has it that genuinely emergent properties, as opposed to "merely" resultant properties, depend on the existence of physical singularities.[12] Otherwise, the sort of asymptotic reasoning required to "see" and understand the properties of interest would fit relatively well the reductive schema (6.1). The novelty of emergent phenomena—their relative unpredictability and inexplicability—is a direct consequence of the failure of reduction. *Thus, the received opinion that part/whole relationships are important for characterizing emergence obscures what is actually most relevant.* Such relationships are operative in some instances, as the thermodynamics of noncritical systems clearly demonstrates, but they are by and large inessential for emergence. What is relevant—that is, what is necessary—is the failure of

---

[12]The adjective "mere" here does not reflect the obvious difficulties of accounting for dominant, *non*emergent phenomena of interest. Much work is required to prove that the thermodynamic limit exists in noncritical systems. Similarly, the mathematics of the interfering ray sum at distances relatively far from the caustic singularity of the ray theory is by no means "mere" or trivial.

the physicists' sense of reduction.   Such reductions fail because of the existence of genuine physical singularities. These singularities can be understood, mathematically, through asymptotic investigations of the kinds discussed. My proposal is that this asymptotic conception of emergence is an improvement over the received view. At the very least, it is a well-defined conception of emergence that applies to many interesting examples in physics for which the received view apparently fails. It accounts for the unpredictable and inexplicable nature of emergents stressed in the received view and *explains* why by appeal to physical singularities and the failure of asymptotic reduction. The account offered by proponents of the received view, to the effect that those aspects—the failures of explainability and of predictability—are the result somehow of wholes being greater than the sum of their parts, misses the point.

8.4   Tenet 5: Novel Causal Powers?

The fifth tenet of the "received" philosophical conception of emergence holds that the novelty of emergent properties consists, in part, in their having causal powers of their own. That is, they have causal powers that are not reducible to the causal powers of their "basal constituents." It is now time to try to determine the extent to which this feature can be accommodated in the new conception of emergence being developed.

As we've just seen, the epistemic aspects of emergence—tenets 3 and 4—result from a failure of the reductive schema (6.1). Can we say the same thing about tenet 5? Do the emergent structures of the rainbow, for instance, possess novel causal powers? In one sense, I think that the answer must be a clear "no." After all, the idea that emergents possess novel causal powers seems to depend in some sense on a contrast between higher level phenomena and their *realizers*. Causal powers of the emergents, if they are novel, are not simply the causal powers of the realizers on any given occasion. But, as we've seen, it is indeed hard to see what are the realizers of those features of the rainbow that we take to be emergent. Such talk seems wholly inappropriate in this context.

At the end of chapter 6 I suggested that the asymptotic borderlands between theories related by singular limits are, in effect, breeding grounds for new theories. Berry has given the name "catastrophe optics" to the theory describing (among other things) the scaling relations for the fringe spacings and intensities of light in the neighborhood of a caustic. I have argued that neither the wave theory nor the ray theory can fully explain these emergent aspects of the behavior of light.[13] Catastrophe optics is an explanatory theory, one that enables us to understand the asymptotically emergent dominant phenomena of interest in the rainbow. In an important sense we cannot understand these features of the rainbow—the behavior captured by a particular instance of equation (6.21)—unless we look to such a novel asymptotic third theory.

As we've seen, in the asymptotic shortwave limit, catastrophe optics sews certain features of the wave theory (the "flesh") onto a skeleton of entities—families

---

[13]See also Batterman (1993, 1995, 1997).

of rays—making up the ontology of the ray theory. The caustic structures of these ray families (and their structural stability) are required to determine the form of the emergent features—the form of the scaling law that describes the phenomenon of interest.

Similarly, we cannot understand the universality of critical phenomena in fluids, for instance, without asymptotically sewing statistical aspects of the behavior of the fluids' components onto singular thermodynamic structures (critical points). These thermodynamic structures are necessary for a complete understanding of the emergent critical phenomena of interest. A third asymptotic theory ("renormalization group" theory), characterizing the singular domain of the thermodynamic limit at critical points, plays a novel explanatory role in this context.

I've also argued in chapter 7, and elsewhere Batterman (1993, 1995, 1997), that semiclassical mechanics plays a similar role as the *theory* of the asymptotic borderland between quantum mechanics and classical mechanics in the semiclassical ($\hbar \to 0$) limit. Most discussions of semiclassical mechanics hold that the semiclassical methods are merely means for approximating solutions to quantum mechanical problems. The main theme throughout this book has been that such asymptotic methods are actually necessary for a proper understanding of certain aspects of (in this case quantum) phenomena of interest. Here, too, wave theoretic or quantum mechanical aspects are sewn onto a skeleton of classical trajectories and their structures in the asymptotic semiclassical limit. This results in a novel explanatory theory as well.

Instead of talking about novel causal powers, I suggest that we reinterpret the fifth tenet of the received view of emergentism as follows:

5. *The explanatory role of the emergents:* Emergent properties figure in novel explanatory stories. These stories involve novel asymptotic theories irreducible to the fundamental theories of the phenomena.[14]

The main reason this reinterpretation is warranted is that it is hard, if not impossible, to see the emergent structures in the rainbow as playing any causal role whatsoever. On the other hand, as I've argued, these structures definitely figure in certain *new* types of asymptotic explanations. They allow us, for instance, to understand the behavior of light near ray theoretic caustics in the shortwave asymptotic limit. But, as far as I can tell, this understanding isn't based upon any causal role played by the wave decorated caustics or by the mathematical characterization of their scaling properties.

Part of the problem I have with the received view is that I just do not understand the talk of causal powers. I realize that largely this is just an appeal to our intuitions about causes, but they are, as far as I can tell, intuitions that resist codification in a coherent theory or framework. It is more fruitful, I think, to adopt the framework of explanation—particularly, asymptotic explanation—as a means to understand the sort of novelty the received view means to capture.

[14] See the discussion in section 8.2.

Many people hold that explanation is causal explanation. If we drop that requirement, we can still refer to the explanations provided by the asymptotic theories as novel. They are distinct in kind, and irreducible to, the explanations that can be provided by the "lower level," "basal," more "fundamental" theory.

8.5 Conclusion

This chapter brings my earlier discussions of the nature of explanation and of reduction to bear on the problem of emergence. Most important, as I've tried to argue, is that we recognize that there is much more to the question of the nature of intertheoretic relations than the philosophical models of reduction would allow. In particular, it is best to think in terms of the physicists' reduction schema and what happens when it fails.

In the event that schema (6.1) fails, we must say that no reduction between the theory pairs (at least in some appropriate regime, e.g., critical phenomena) can obtain. I take this to be a necessary condition for the emergence of novel structures or properties. These structures or properties will inhabit the asymptotic "no man's land" between the theories. I've argued that they will possess features that quite reasonably should lead us to think of them as emergent structures or properties.

This view entails that it is fruitful to think of emergence in terms of a pair of theories and their relationships. Nothing like this is explicit in the "received view" of emergence, although I think it is definitely implicit. After all, the very notions of predictability and explainability are theory-relative. These epistemological notions depend on metaphysical or ontological relations between theories—relations that require a sufficient mathematization of the theories in order to be seen.

We have also seen that the idea of emergence as essentially involving some conception of the properties of wholes transcending the properties of their parts, is often a red herring. Instead, the crucial feature is the singular nature of the limiting relationship between the theory pairs. Of course, this is not, as we've seen, to say that no part/whole relations will play a role in the formation and description of certain properties of wholes. It's just that without singular limiting relations the properties of the wholes will be resultant and not emergent.

The view of emergence presented here depends upon the fact, just noted, that the theories involved are expressible in a sufficiently rich mathematical language so that the asymptotic relations between them can be explored. This may lead many to wonder about the relevance of the view to the context in which most philosophical discussions take place, namely, the context of mental phenomena. My own view about this is that we need to wait and see how various research programs in neuropsychology and neurophysiology play out. Perhaps, in the near future sufficiently sophisticated theories will be forthcoming and the kind of analysis I've been discussing here will be possible. Of course, one problem—and a deep one at that—is the nature of the "upper level" theory in this context. Folk psychology, by its very nature, will likely never be expressible in a sufficiently formal language for this analysis to apply. Nevertheless, to me, it doesn't seem unreasonable to expect that the emergence of certain mental phenomena, if they are genuinely emergent, will depend on some sort of limit—for instance, the limit of increasing complexity.

9

Conclusions

This book has been investigating some aspects of asymptotic reasoning in science. In broad strokes, such reasoning involves the idea that sometimes throwing away certain types of details and giving up on certain levels of precision will lead to a proper (correct) understanding of a given phenomenon. In one sense this should not be too surprising. Isn't this just what is often involved in the process of idealization? Surely, no one would deny the importance of idealization in scientific theorizing. The problem, however, is to make sense of this from the point of view of physical theory itself. If we have a well-confirmed theory that provides detailed information about the phenomenon, then it seems that, in principle, having those details ought to be our best bet at fully understanding the phenomenon of interest. However, I have been arguing that in many situations this is simply not the case.

In introducing the unification model of explanation, Michael Friedman claims that while most of the philosophical literature on explanation deals with the explanation of particular events, such explanations are comparatively rare in science itself. Most often, he claims, "what is explained is a general regularity or pattern of behavior—a law, if you like" (Friedman, 1974, p. 5). There is much truth to this claim. As a result of investigations in the pages here, we can see what is responsible for this disparity in models of explanation. There is an important ambiguity involved in speaking of explaining or understanding some physical phenomenon. By "phenomenon," do we mean the particular event or some pattern or regularity that subsumes it?

No one would deny that if I want to understand the buckling of a strut, I am trying to account for some phenomenon that I find to be of interest. Clearly, pragmatic considerations are at play here. Suppose that the reason I find this interesting is because I've just witnessed the buckling. The next step, almost certainly, is to try to repeat the event. Is this something that happens on a regular basis once certain conditions are met? If not, then while I may still search for an explanation of what I've just witnessed, it has lost some of its interest, since I will not be able to project anything about what I've witnessed onto other similar situations. The explanation will be pragmatically useless: I won't advert to it when I'm trying to figure out if my building or bridge will collapse under certain loads.

If, on the other hand, I am able to reproduce the results in other struts, then I have much more reason to try to understand what is going on. There is still the question of why each particular buckling event occurs, and it is upon this that the philosophical accounts of explanation Friedman addresses focus. They often ignore the more fundamental question about reproducible phenomena; namely, why does the regularity or pattern itself obtain.[1] In section 3.2 we have seen reasons to be sceptical about the unification theorists' own answer to this question. I will not rehearse these reasons here. The result of that discussion, together with the my discussion of the failures of causal-mechanical models, is that the proper focus should be on explaining the universality highlighted by the fact that the regularity exists. Given the detailed differences that necessarily are operative in each particular case we observe, what makes the reproducibility or repeatability of the "phenomenon" possible? How can we account for the universality in the pattern of behavior?

When an applied mathematician sets out to "model" some, even moderately complex, phenomenon, she looks to answer just this question. The details of any given instance of the reproducible phenomenon are simply too much to deal with. Besides, the real interest, as I've tried to argue, is in the existence of the observed pattern itself. The applied mathematician seeks to *simplify or idealize* the problem so as to focus on those factors essential to the universal behavior observed. Barenblatt, as the reader will recall, refers to these as the "basic" or "principal" features of the phenomenon. Asymptotic reasoning—the taking of limits as a means to simplify, and the study of the nature of these limits—constitutes the main method of idealization in the mathematician's tool box.

As we have seen, genuine insight is often the result of such idealizing asymptotic investigations. It is *not* always true that the results—the principal features or "laws"—arrived at by these methods are *merely* the first steps toward a deeper understanding to be obtained when (and if) we make further strides in computation. There is a trade-off between "improving" upon this (asymptotic) type of idealized result, by including more details in the computations, and achieving the understanding of patterns one desires. Such "improvements" move one away from an account of the universality of the behavior, toward explanations of particular instances of that universal behavior. The type of why-question shifts and the question about the existence of the pattern becomes less salient.

As I've said, I think Friedman is right that most philosophical accounts of explanation focus on the explanation of particular events. They do so because they are misled by an inaccurate and somewhat underdeveloped understanding of the role of idealization in science. Their motto is something like this: "Of course, as a first step, to get the ball rolling so to speak, we need to idealize,

---

[1] More accurately, they just assume it is answered by appeal to a misguided ideal—"fundamental theory."

but for true explanation and understanding, we must remove that idealization (or 'improve' upon it) by including more and more details."

One of the most important lessons to be learned from the examples discussed in this book is that this attitude is naive and often misguided. In the context of explanation, the success of asymptotic investigations in providing scientific understanding, and genuine answers to important explanatory why-questions, testifies to the fact that such "improvements" are by no means necessary or even desirable.

In the context of reduction, or more properly, of intertheoretic relations, we have seen that asymptotic methods also play an extraordinarily important role. The modern history of the development and evolution of scientific theories is replete with examples in which one theory supersedes another. In the process the newer theory (T_f) often preserves, to some extent, the domain of the previous one (T_c), while at the same time expanding to encompass aspects that were not treated by the earlier theory. To a large extent these relations between the earlier and later theories are captured in the mathematical formalism.[2] With hindsight, much can be learned by studying how the formalisms of the two theories are related.

However, investigations of this sort are not purely mathematical. As we've seen repeatedly throughout the book, physical intuition plays a crucial role in interpreting the results of such investigations. Recall, for example, the discussion of the construction of the interfering ray sum in section 6.3. The most surprising conclusion I have drawn concerns the autonomy and independence (in some sense) of the limiting domain between theories related by *singular* asymptotic limits. Theory pairs for which the physicists' reductive schema

$$\lim_{\epsilon \to 0} T_f = T_c \qquad (9.1)$$

fails, will typically be accompanied by new phenomena irreducible either to T_f or T_c. Third theories of this asymptotic domain figure essentially in explanatory accounts of the nature of the new phenomena.

Discussions in the philosophical literature on reduction largely ignore the limiting (mathematical) aspects of the intertheoretic relations. Instead, they assume that since the newer theory T_f purports to improve upon and account for the (partial) successes of the earlier theory, there must be relations among the predicates of the two theories that can be expressed in the form of bridge laws. These types of semantic connections, if they exist at all, are often difficult to formulate and difficult to justify. But in any case, they won't serve to aid in the understanding of the new and surprising features that appear in the asymptotic borderland between the theories.

Recently, a major focus of the philosophical reduction literature has been on the status of the special sciences. It is difficult to subject these theories to

---

[2] It is important to emphasize that not all aspects of intertheoretic relations will be captured by the differences in mathematical formalisms. We need only think of the difficulties in "reducing" thermodynamics to statistical mechanics to see this point. For instance, in this case there are deep puzzles about the introduction and interpretation of probabilities in statistical mechanics.

the same sort of rigorous asymptotic investigations as was possible in the case of the wave and ray theories of light, in the limiting relations between quantum and classical mechanics, and in the case of thermodynamics and statistical mechanics. The reason is obvious. The upper level theories of the special sciences are, in general, not yet sufficiently formalized (or formalizable) in an appropriate mathematical language. Despite this limitation this discussion has made considerable progress in sharpening the issues involved in the debate.

We have seen that the important questions concern the possibility of explaining the universality of the upper level properties and generalizations from the point of view of the lower level theory. The defense of a robust nonreductive physicalism requires that we explain this universality while simultaneously allowing for the failure of a philosophical reduction of the upper level theory to the lower level, physical theory. Asymptotic explanation of this universality, it seems to me, is our best bet here. The multiple realizability argument invokes the wildly heterogeneous nature of the lower level realizers to block reduction, where reduction is understood in terms of the availability of bridge laws relating natural kind properties at the upper level to natural kinds at the level of the realizers. If reduction is understood in this way, then I have argued that it is reasonable to hold that reduction will fail.

On the other hand, on most philosophical conceptions of reduction, to reduce is to explain. Reduction, that is, is taken to be explanatory reduction. So, if reduction fails because of multiple realizability, then no explanation will be available. This puts severe pressure on the nonreductive physicalist. In what sense is her physicalism at all interesting and robust? At best, it seems we have what Fodor has called token-token identities. Particular upper level property instances are identified with particular lower level property instances. But then we lose any type of understanding of the universality of the upper level features. What unifies all of the diverse token-token relations? Kim, as we've seen, exploits this arguing in effect that, to maintain an interesting physicalism, either we must give up the position that the special sciences are genuine sciences, or we must buy into an extreme reductionist position—a reductionism that entails that at best there will be only species- or structure-specific special sciences.

Our attention to the nature of asymptotic explanation holds out the possibility that we can explain the universality of the special sciences from the point of view of the lower level theory while maintaining the irreducibility of those sciences to physics. Explanation and reduction must part company. The philosophical model that weds the two is simply mistaken. I have tried to argue that is not unreasonable to think that such asymptotic methods will play a role in our understanding of the status of the special sciences when, and if, they become sufficiently mathematized. It is hard, indeed, to see how the explanation of universality (or multiple realizability) would proceed without the discovery of principled reasons grounded in the theory of the realizers, to the effect that most of the details distinguishing those realizers are *irrelevant* and must be ignored in the context of understanding that universality. Once one recognizes that a claim of multiple realizability is just a statement of universality, the asymptotic methods developed by the physicists become obvious candidates for dealing with

Conclusions

the problem of the special sciences.

Clearly, much more work needs to be done in this context. It is my hope that this discussion will lead investigators with more knowledge of the special sciences than I to consider asymptotic methods in their investigations into the status of those sciences.

Kim has also argued that nonreductive physicalism is modern emergentism. In some sense I believe he is right about this. The problem, however, is to understand properly the nature of emergence.

Our investigation into the singular limits that often exist between theories has led us to examine universal phenomena that appear to emerge in asymptotic domains. I've argued that the use of the term "emergent" to describe the structures and features of these phenomena is wholly appropriate. They are (relative to the various theory pairs involved) genuinely novel structures. In the physics literature one often finds claims to the effect that these phenomena constitute "new physics" requiring novel theoretical work—new asymptotic theories—for their understanding.

In keeping with the "received view" of emergence to be found in the philosophical literature, I've argued that emergent phenomena are necessarily irreducible in the strongest possible sense. This strong sense of irreducibility is expressed by the failure of the reductive schema (9.1). Such reductive failure will entail the failure of any sort of philosophically motivated reduction as well. However, in contrast to the received view, I hold that this irreducibility is consistent with the possibility that the existence and nature of the emergents are susceptible to asymptotic explanatory scrutiny. The phenomena are not brute and inexplicable *tout court*. They are brute in the sense that they are irreducible to, unpredictable from, and unexplainable from the "base" theory T_f alone—that is, without the use of asymptotic methods. Our recognition that the connection between reduction and explanation must be severed allows for this somewhat weakened sense of emergence.

The focus on the failure of schema (9.1) as at least a necessary condition for emergence allows us to see that the received view inappropriately highlights mereological criteria in its expression of the emergentist position. There are examples—the rainbow being one—in which reduction in the strong sense fails, but where no part/whole relations obtain.

Here, too, it is clear that more work needs to be done. Those wedded to a conception of emergence as involving inexplicably brute phenomena will likely not be overly impressed by the new view I advocate. But, then, I think it is incumbent upon these investigators to provide us with genuine examples of emergent properties. If it is reasonable to think of nonreductive physicalism (modified to allow for explanation of the universal, multiply realized properties and regularities) as an expression of the emergentist doctrine, then these "genuinely" emergent properties will have to be very strange indeed.

Bibliography

Airy, G. B. 1838. On the intensity of light in the neighborhood of a caustic. *Transactions of the Cambridge Philosophical Society*, 6:379–403.

Andronov, A., and Pontryagin, L. 1937. Systèmes grossiers. *Dokl. Akad. Nauk. SSSR*, 14:247–251.

Antony, Louise. 1999. Making room for the mental: Comments on Kim's "Making sense of emergence." *Philosophical Studies*, 95(1/2):37–44.

Arnold, V. I. 1989. *Mathematical Methods of Classical Mechanics*. New York: Springer-Verlag, 2nd edition. Trans. Vogtmann, K. and Weinstein, A.

Barenblatt, G. I. 1996. *Scaling, Self-similarity, and Intermediate Asymptotics*. Cambridge Texts in Applied Mathematics. Cambridge: Cambridge University Press.

Batterman, R. W. 1991. Chaos, quantization, and the correspondence principle. *Synthese*, 89:189–227.

Batterman, R. W. 1992. Explanatory instability. *Nous*, 26(3):325–348.

Batterman, R. W. 1993. Quantum chaos and semiclassical mechanics. In *PSA 1992*, (vol. 2, pp. 50–65). East Lansing, MI: Philosophy of Science Association.

Batterman, R. W. 1995. Theories between theories: Asymptotic limiting intertheoretic relations. *Synthese*, 103:171–201.

Batterman, R. W. 1997. "Into a mist": Asymptotic theories on a caustic. *Studies in the History and Philosophy of Modern Physics*, 28(3):395–413.

Batterman, R. W. 1998. Why equilibrium statistical mechanics works: Universality and the renormalization group. *Philosophy of Science*, 65:183–208.

Batterman, R. W. 2000. Multiple realizability and universality. *British Journal for the Philosophy of Science*, 51:115–145.

Bender, C. M., and Orszag, S. A. 1978. *Advanced Mathematical Methods for Scientists and Engineers*. New York: McGraw-Hill.

Berry, M. V. 1981. Singularities in waves and rays. In Balian, R., Kléman, M., and Poirier, J. P., (Eds.), *Physics of Defects*, Les Houches, Session 35, 1980, (pp. 453–543). Amsterdam: North-Holland.

Berry, M. V. 1983. Semiclassical motion of regular and irregular motion. In Iooss, G., Helleman, R. H. G., and Stora, R., (Eds.), *Chaotic Behavior of Deterministic Systems*, Les Houches Session 36, 1981, (pp. 171–271). Amsterdam: North-Holland.

Berry, M. V. 1987. The Bakerian Lecture. quantum chaology. In Berry, M. V., Percival, I. C., and Weiss, N. O., (Eds.), *Dynamical Chaos*, (vol. 186, pp. 183–198). Princeton: Royal Society of London, Princeton University Press.

Berry, M. V. 1990. Beyond rainbows. *Current Science*, 59(21 & 22):1175–1191.

Berry, M. V. 1994. Asymptotics, singularities and the reduction of theories. In Prawitz, Dag, Skyrms, Brian, and Westerståhl, Dag, (Eds.), *Logic, Methodology and Philosophy of Science, IX: Proceedings of the Ninth International Congress of Logic, Methodology and Philosophy of Science, Uppsala, Sweden, August 7-14, 1991*, vol. 134 of *Studies in Logic and Foundations of Mathematics*, pp. 597–607. Amsterdam: Elsevier Science B. V.

Berry, M. V., and Upstill, C. 1980. Catastrophe optics: Morphologies of caustics and their diffraction patterns. In Wolf, E., (Ed.), *Progress in Optics*, (vol. 18, pp. 257–346). Amsterdam: North-Holland.

Block, Ned. 1997. Anti-reductionism slaps back. *Philosophical Perspectives*, 11: 107–132.

Chen, L-Y, Goldenfeld, N., Oono, Y., and Paquette, G. 1994. Selection, stability and renormalization. *Physica A*, 204:111–133.

Churchland, P. S., and Sejnowski, T. J. 1992. *The Computational Brain*. Cambridge, MA: MIT Press.

Feyerabend, P. K. 1962. Explanation, reduction, and empiricism. In Feigl, H. and Maxwell, G., (Eds.), *Minnesota Studies in the Philosophy of Science*, (vol. 3, pp. 28–97). Minneapolis, University of Minnesota Press.

Fisher, Michael E. 1983. Scaling, universality and renormalization group theory. In Hahne, F.J.W., (Ed.), *Critical Phenomena*, vol. 186 of *Lecture Notes in Physics*. Berlin: Springer-Verlag.

Fodor, Jerry. 1974. Special sciences, or the disunity of sciences as a working hypothesis. *Synthese*, 28:97–115.

Fodor, Jerry. 1997. Special sciences: Still autonomous after all these years. *Philosophical Perspectives*, 11:149–163.

Fowler, A. C. 1997. *Mathematical Models in the Applied Sciences*. Cambridge Texts in Applied Mathematics. Cambridge: Cambridge University Press.

Friedman, Michael. 1974. Explanation and scientific understanding. *Journal of Philosophy*, 71(1):5–19.

Goldenfeld, N., Martin, O., and Oono, Y. 1989. Intermediate asymptotics and renormalization group theory. *Journal of Scientific Computing*, 4(4):355–372.

Greenler, Robert. 1980. *Rainbows, Halos, and Glories*. Cambridge: Cambridge University Press.

Guckenheimer, John and Holmes, Philip. 1983. *Nonlinear Oscillations, Dynamical Systems, and Bifurcations of Vector Fields*, vol. 42 of *Applied Mathematical Sciences*. New York: Springer-Verlag.

Gutzwiller, Martin C. 1990. *Chaos in Classical and Quantum Mechanics*. New York: Springer-Verlag.

Heller, Eric J., and Tomsovic, Steven. 1993. Postmodern quantum mechanics. *Physics Today*, 46:38–46.

Hempel, Carl G. 1965. Aspects of scientific explanation. In *Aspects of Scientific Explanation and Other Essays in the Philosophy of Science*, (pp. 331–496). New York: The Free Press.

Humphreys, Paul. 1997. How properties emerge. *Philosophy of Science*, 64(1): 1–17.

Jammer, Max. 1966. *The Conceptual Development of Quantum Mechanics*. New York: McGraw-Hill.

Jones, Todd. 1995. Reductionism and the unification theory of explanation. *Philosophy of Science*, 62:21–30.

Khinchin, A. I. 1949. *Mathematical Foundations of Statistical Mechanics*. New York: Dover Publications. Translation: G. Gamow.

Kim, Jaegwon. 1992. Multiple realization and the metaphysics of reduction. *Philosophy and Phenomenological Research*, 52(1):1–26.

Kim, Jaegwon. 1994. Explanatory knowledge and metaphysical dependence. In *Truth and Rationality*, Philosophical Issues, 5, pp. 52–69.

Kim, Jaegwon. 1998. *Mind in a Physical World: An Essay on the Mind-Body Problem and Mental Causation*. Cambridge, MA: The MIT Press.

Kim, Jaegwon. 1999. Making sense of emergence. *Philosophical Studies*, 95 (1/2):3–36.

Kitcher, Philip. 1989. Explanatory unification and the causal structure of the world. In Kitcher, Philip and Salmon, Wesley C., (Eds.), *Scientific Explanation*, vol. 13 of *Minnesota Studies in the Philosophy of Science*, pp. 410–505. Minneapolis: University of Minnesota Press.

Littlejohn, Robert G. 1992. The van Vleck formula, Maslov theory, and phase space geometry. *Journal of Statistical Physics*, 68(1/2):7–50.

Mazur, P., and van der Linden, J. 1963. Asymptotic form of the structure function for real systems. *Journal of Mathematical Physics*, 4(2):271–277.

Nagel, Ernest. 1961. *The Structure of Science*. London: Routledge and Kegan Paul.

Nickles, Thomas. 1973. Two concepts of intertheoretic reduction. *Journal of Philosophy*, 70(7):181–201.

Papineau, David. 1993. *Philosophical Naturalism*. Oxford: Blackwell.

Pfeuty, P., and Toulouse, G. 1977. *Introduction to the Renormalization Group and to Critical Phenomena*. London: John Wiley and Sons. Translation: G. Barton.

Piexoto, M. 1962. Structural stability on 2-dimensional manifolds. *Topology*, 1: 101–120.

Poston, T., and Stewart, I. 1978. *Catastrophe Theory and its Applications*. London: Pitman.

Putnam, H. 1980. The meaning of 'meaning'. In *Mind, Language and Reality: Philosophical Papers*, (vol. 2, pp. 215–271). Cambridge: Cambridge University Press.

Railton, Peter. 1981. Probability, explanation, and information. *Synthese*, 48: 233–256.

Rohrlich, Fritz. 1988. Pluralistic ontology and theory reduction in the physical sciences. *British Journal for the Philosophy of Science*, 39:295–312.

Salmon, Wesley C. 1989. Four decades of scientific explanation. In Kitcher, Philip and Salmon, Wesley C., (Eds.), *Scientific Explanation*, vol. 13 of *Minnesota Studies in the Philosophy of Science*, pp. 3–219. Minneapolis: University of Minnesota Press.

Schaffner, Kenneth. 1976. Reductionism in biology: Prospects and problems. In Cohen, R. S. et al., (Eds.), *PSA 1974*, (pp. 613–632). Boston: D. Reidel Publishing Company.

Schurz, G., and Lambert, K. 1994. Outline of a theory of scientific understanding. *Synthese*, 101:65–120.

Sklar, Lawrence. 1967. Types of inter-theoretic reduction. *British Journal for the Philosophy of Science*, 18:109–124.

Sklar, Lawrence. 1993. *Physics and Chance: Philosophical Issues in the Foundations of Statstical Mechanics*. Cambridge: Cambridge University Press.

Smale, S. 1980. What is global analysis? In *The Mathematics of Time: Essays on Dynamical Systems, Economic Processes, and Related Topics*, (pp. 84–89). New York: Springer-Verlag.

Teller, Paul. 1992. A contemporary look at emergence. In Beckermann, A., Flohr, H., and Kim, J., (Eds.), *Emergence or Reduction? Essays on the Prospects of Nonreductive Physicalism*, Foundations of Communication and Cognition, (pp. 139–153). Berlin: Walter de Gruyter.

Wimsatt, William. 1994. The ontology of complex systems: Levels, perspectives, and causal thickets. *Canadian Journal of Philosophy*, Supplementary Volume 20:207–274.

Woodward, James. 2000. Explanation and invariance in the special sciences. *British Journal for the Philosophy of Science*, 51:197–254.

Index

Airy integral, 81, 83, 84, 92n, 93
Airy, G. B., 83
Andronov, A., 58n
Antony, L., 115n
Arnold, V. I., 102n
asymptotic borderland, 6, 77, 99,
    110, 111, 116
    and emergence, 95, 121
    and reduction, 95, 133
    theory of, 6, 77, 119, 126
asymptotic reasoning, 3, 7, 13, 19,
    23, 30, 37, 46, 49, 52, 56,
    57, 76, 96, 124, 131
    and universality, 16

Barenblatt, G. I., 15, 16, 44–46, 46n,
    51–53, 55, 55n, 57, 78, 94,
    124, 132
Batterman, R. W., 23, 23n, 29, 30n,
    34, 35n, 56, 74, 75n, 77,
    78, 79n, 85n, 90n, 100, 102n,
    105n, 107n, 110n, 123n, 126n,
    127
Bender, C. M., 79, 90
Berry, M. V., 13, 77, 80n, 84, 84n,
    88, 89n, 92–96, 102n, 107,
    110, 116, 117n, 125, 126
Block, N., 61, 71, 72, 72n, 75, 76
Bohr, N., 99, 100, 107, 109

catastrophe optics, 77, 84–93, 106
    as theory, 95, 97
catastrophes, 91, 92, 100
    and caustics, 91
    diffraction integrals, 92, 93, 108
        scaling laws, 93–94
    normal forms, 92

    theory of, 85, 91
caustics, 81–84, 101, 107
    and Airy integral, 84
    and catastrophes, 91, 92, 126
    and chaos, 109
    and emergence, 113, 115, 120,
        127
    and interfering ray sum, 88, 96
    and Maslov's method, 90, 108
    and ray theory, 82, 84, 117
    and stability, 91
    and wave theory, 84
chaos, 109
chaos game, 23–25, 28, 29
Chen, L-Y, 58, 59
Churchland, P. S., 74
correspondence limit, 5, 18, 78, 99,
    109, 111
correspondence principle, 5, 99
critical phenomena, 4, 12, 13, 35,
    37–42, 75, 80, 93, 121, 124,
    125, 127, 128

Descartes, 82
dimensional analysis, 15–16, 19, 44,
    48–50, 52, 55, 56
    and intermediate asymptotics,
        51, 52, 55
    as asymptotic reasoning, 49
Disney principle, 72

emergence, 6, 19–22, 97
    and asymptotics, 74, 126, 128
    and nonreductive physicalism,
        135
    and parts/wholes, 20, 21, 113,
        118, 125, 128, 135

emergence (*continued*)
    and reduction, 6, 121, 124, 126, 135
    and the rainbow, 113, 115–120
    the "received view" of, 20, 115, 135
Euler formula, 12, 13
Euler Strut, 9–13
explanation, 133
    and emergence, 118
    and ideal texts, 10
    and multiple realizability, 72, 74–76, 119, *see* universality
    and reduction, 6, 61, 95, 115, 134
    and stability, 57–59
    and the renormalization group, 37
    and understanding, 12, 56
    and why-questions, 23–25, 31, 70, 132
    asymptotic, 4, 37, 74, 118, 127
    causal-mechanical, 10
    D-N, 26
    D-N-P, 28
    I-S, 26
    of critical phenomena, 37–42
    of universality, 42–44
    unification model, 30–35, 131

Feyerabend, P. K., 64
Fisher, M. E., 12, 35, 45, 90
Fodor, J., 61, 65–68, 70–73, 75, 76, 134
Fowler, A. C., 52, 56
Friedman, M., 30, 32, 131, 132
fringe spacings, 88, 90, 93–96
    universality of, 93, 94
functionalism, 68

generating function, 85n, 90, 92, 92n, 93, 94, 103
Goldenfeld, N., 43, 44, 48, 53, 55
Greenler, R., 82n
Guckenheimer, J., 58n
Gutzwiller, M., 110, 110n

Heller, E. J., 110n
Helmholtz equation, 86, 96, 104, 118
Hempel, C. G., 25, 26, 26n, 27, 29
Holmes, P., 58n
Humphreys, P., 20, 114n

interfering ray sum, 88–90, 93, 102–106, 117, 118, 125, 133
    and caustics, 88
    and scaling laws, 96
    and stationary phase, 90
    failure of, 88, 90
intermediate asymptotics, 44–46, 78, 80, *see* dimensional analysis
    and renormalization group, 55
    first kind, 46–52
    second kind, 52–57

Jammer, M., 99
Jones, T., 34n

Khinchin, A. I., 122, 123n
Kim, J., 20, 21, 21n, 32, 33, 61, 65, 65n, 66–70, 70n, 71–73, 75, 76, 113n, 114, 114n, 115, 118, 120, 125, 134, 135
Kitcher, P., 30–34

Lambert, K., 30
limit
    semiclassical, 99, 102, 106, 107, 109–111, 127
    shortwave, 81, 84, 88, 93, 95, 99, 106, 116, 117, 120, 126, 127
    singular, 5, 6, 16, 18, 21, 49n, 51, 53, 56, 78, 80–81, 94–96, 99, 103, 109, 110, 117n, 124–126, 128, 133, 135
    new physics, 16, 78, 96
    singular vs. regular, 18–19
    thermodynamic, 122–123
Littlejohn, R. G., 102

Maslov's method, 90, 108
Mazur, P., 123

Index

multiple realizability, 4, 34n, 61, 65–68, 72, 119, 134
    and universality, 73–76

Nagel, E., 61–63
Nickles, T., 5, 18n, 64, 78, 78n

optical distance function, 85, 86, 90, 92, 93, *see also* generating function
order parameter, 39
Orszag, S. A., 79, 90

Papineau, D., 72
Pfeuty, P., 40
physicalism, 73, 76, 134
    and emergence, 21, 135
Piexoto, M., 58n
Pontryagin, L., 58n
Poston, T., 91n
Putnam, H., 65n, 68

quantum chaos, 109

Railton, P., 10, 11, 25, 27–29, 30n, 33
rainbow, 5, 81–84
    and catastrophe optics, 126
    and emergence, 21, 113, 115–120
    and reduction, 77, 95
    caustic, 82
ray theory, 5, 17, 77, 93
    and caustics, 82, 84, 94
reduction, 5, 17–19, 77, 133
    and asymptotics, 76
    and emergence, 6, 19, 120, 121, 125, 126, 128
    and explanation, 6, 115, 134
    and multiple realizability, 65–66, 134
    as "simplification", 52, 56
    functional model, 61, 67–71
    Nagelian, 61–65
    philosophers' sense, 5, 95, 117
    physicists' sense, 5, 65, 78–80, 117, 124

renormalization group, 4, 35, 37–43, 74, 75, 80
    and intermediate asymptotics, 53, 55, 56
    features of, 43
Rohrlich, F., 79

Salmon, W., 10, 30, 31
scaling, *see* self-similarity
Schaffner, K., 63–65, 79
Schrödinger equation, 12, 39, 96, 106, 107, 110
    and WKB method, 104–106
Schurz, G., 30
Sejnowski, T. J., 74
self-similarity, 41, 48–50, 52, 53, 55, 56, 93, 94, *see also* intermediate asymptotics
    importance of, 51
semiclassical approximation, 88
semiclassical mechanics, 95, 100, 107, 119n
    as theory, 99, 100, 127
Sklar, L., 26n, 62, 63, 63n, 121, 123n
Smale, S., 58n
special sciences, 4, 65, 68, 71, 74, 114
    and emergence, 113, 113n
    and multiple realizability, 65, 67, 72, 75
    and physicalism, 73
    and reduction, 61, 65, 68, 76, 133
    and universality, 134, 135
stability, 34, 42, 44, 51, 57–59, 91, 93, 94, 127
stationary phase, 90
    method of, 90, 93, 118
Stewart, I., 91n
supervenience, 69n, 114

Teller, P., 20, 118n
Thom, R., 91
Tomsovic, S., 110n
torus quantization, 107
    and chaos, 107, 109

Toulouse, G., 40
turning points
    as caustics, 106

understanding, 3, 25, 30, 31, 90, 96, 131, 133
    and asymptotics, 46, 52, 96, 104, 121, 127, 131–133, 135
    and explanation, 10, 12, 56
    and stability, 59
    and unification, 30–32
    causal-mechanical, 30
    of universality, 34, 45, 46
universality, 4, 6, 9, 13, 14, 17, 23, 37, 39, 71, 93–94, 110, 119, 132
    and multiple realizability, 61, 73–76
    and why-questions, 23
    explanation of, 42–44, *see also* renormalization group
    features of, 13, 38, 73
    of critical phenomena, 37, 42, 53
    relation to macroscopic phenomenology, 43
Upstill, C., 77, 84, 89n, 92–94

van der Linden, J., 123

wave theory, 5, 6, 17, 21, 77, 80, 83, 93, 95
why-questions, 4, 4n, 132, 133
    and unification, 33
    types (i) and (ii), 23, 31, 70
Wimsatt, W., 21n, 64
WKB method, 104–109, *see also* semi-classical mechanics and semi-classical approximation
Woodward, J., 59n

Young's modulus, 12, 12n, 14